Second Generation Wavelets and Applications

Maarten Jansen and Patrick Oonincx

Second Generation Wavelets and Applications

With 58 Figures

Springer

Maarten Jansen, MSc Engineering, PhD Computer Science
Department of Mathematics and Computer Science
TU Eindhoven, PO Box 513
5600 MB Eindhoven, The Netherlands
and
Department of Computer Science
K.U. Leuven, Celestijnenlaan 200A
3001 Leuven, Belgium

Patrick Oonincx, MSc Mathematics, PhD Mathematics
Royal Netherlands Naval College
PO Box 10000
1780 CA Den Helder, The Netherlands

British Library Cataloguing in Publication Data
Jansen, Maarten
  Second generation wavelets and applications
  1. Signal processing — Mathematics   2. Image processing —
  Mathematics   3. Wavelets (Mathematics)
  I. Title   II. Oonincx, Patrick
  515.2′433
ISBN 1852339160

Library of Congress Cataloging-in-Publication Data
Jansen, Maarten.
   Second generation wavelets and applications / Maarten Jansen and Patrick Oonincx.
      p. cm.
   Includes bibliographical references and index.
   ISBN 1-85233-916-0 (alk. paper)
   1. Signal processing—Mathematics. 2. Image processing—Mathematics. 3. Wavelets
(Mathematics)   I. Oonincx, Patrick. II. Title.
   TK5102.9.J3623 2005
   515′.2433—dc22   2004058967

Apart from any fair dealing for the purposes of research or private study, or criticism or review, as permitted under the Copyright, Designs and Patents Act 1988, this publication may only be reproduced, stored or transmitted, in any form or by any means, with the prior permission in writing of the publishers, or in the case of reprographic reproduction in accordance with the terms of licences issued by the Copyright Licensing Agency. Enquiries concerning reproduction outside those terms should be sent to the publishers.

ISBN 1-85233-916-0
Springer Science+Business Media
springeronline.com

© Springer-Verlag London Limited 2005

The use of registered names, trademarks, etc. in this publication does not imply, even in the absence of a specific statement, that such names are exempt from the relevant laws and regulations and therefore free for general use.

The publisher makes no representation, express or implied, with regard to the accuracy of the information contained in this book and cannot accept any legal responsibility or liability for any errors or omissions that may be made.

Typesetting: Electronic text files prepared by authors
Printed in the United States of America
69/3830-543210   Printed on acid-free paper   SPIN 10974018

Preface

The ongoing success of wavelet theory and similar multiscale data decompositions is largely due to its flexibility in continuously expanding its horizon towards new applications. The first success of wavelets came in the 1980s with the continuous wavelet transform as a method in the time-scale analysis of signals. While this research went on (and still goes on), it was soon realized that the concept of time-scale analysis is closely related to the sort of signal processing that happened in so-called Laplacian pyramids. The formal definition of multiresolution, the corresponding fast wavelet transform algorithm (Mallat) and the discovery of compactly supported orthogonal wavelets (Daubechies) were major breakthroughs in the development of the discrete wavelet transform as a tool in signal and image processing. In spite of this immediate success, the classical discrete wavelet transform is somehow limited. A typical example of such a limitation is the assumption that the input has to be a regularly observed (sampled) signal, where the number of observations is a power of two. Obviously, these limitations can be dealt with by some pre- and/or post-processing of the data: think about interpolation for irregularity and symmetric extension if the number of data is not a power of two. Nevertheless, it seems more interesting if we can incorporate the grid structure and the interval boundaries into the actual construction of the multiresolution analysis. This is exactly what was aimed by 'second-generation wavelets' when they were first proposed (Sweldens), about 10 years ago.

The construction of second-generation wavelets is based on the lifting scheme. In the first instance, this scheme is an alternative implementation of the fast discrete wavelet transform. It is a bit faster than the classical repeated filterbank implementation and it allows in-place computations, i.e., no additional working memory is needed, as all steps in the algorithm may overwrite input data without making the transform irreversible. It was soon realized that this lifting scheme has another, important advantage: unlike the classical filterbank implementation, the concept is readily extensible to differently structured data. This includes not only the above-mentioned irregularly observed data and data on an interval, but nearly any type of structure. One could now think, for instance, about a multiresolution analysis of DNA-molecule structures, large networks, surfaces in computer graphics applications and so on. Even for common images, it pays off using the lifting scheme, as was illustrated by the new compression standard in JPEG-2000, which makes use of lifting. Lifting also allows one to add a multiscale element to previously developed methods. In statistics, for instance, lifting can be used in combination with existing smoothing methods, such as spline smoothing.

Another extension, equally easy to construct with the lifting scheme, are nonlinear and data-adaptive multiscale analyses. A well known example of a nonlinear multiresolution decomposition is the integer wavelet transform that maps an array of integers to another array of integers. Data-adaptive multiresolution transforms are extensively discussed further in this book. In two dimensions, a specific data-adaptive, anisotropic approach, discussed in the last chapter, shows interesting behaviour in the neighbourhood of edges in images. While the concept of this edge-adaptive multiscale decomposition is quite easy to understand, the resulting convergence rate compares to that of more complicated approaches, such as curvelets, contourlets, edgelets and so on.

Wavelet theory and applications are situated on the edges between many fields: signal and image processing, numerical analysis and scientific computing, statistics, approximation theory and other mathematical fields. Each of these domains has its own terminology and requires a specific writing style. While people in signal processing are used to illustrations and diagrams, mathematicians prefer clear formulas to think and reason about. Samples in signal processing correspond to observations in statistics, while a sample in statistical literature is mostly the entire set of observations. Signals in signal processing are functions in mathematics. We have tried to offer both figures and formulas, fully aware that some readers might spend some time on puzzling out the meaning of each symbol in an expression or a diagram. Yet, we believe that doing so is a key to opening the doors of many papers that otherwise remain unread. Also, we tried to provide sufficient variation, in the sense that a mathematical approach is illustrated with figures and plots in a subsequent section and vice versa.

As for the mathematical depth of our approach, our point of view is to provide a guide for the exploration of the literature, rather than a fully elaborated mathematical treatment of all aspects of second-generation wavelets. We therefore omit some of the full, rigourous proofs if they would take too much space and attention. We hope to compose a clear general overview on the topic of second-generation wavelets.

Writing and even thinking this book would be a much harder job without the presence of our families.

I, Maarten Jansen, would like to express my gratitude to my wife, Gerda Claeskens, of course for her support and patience, but also for her keen curiosity and interest in my work. Her remarks after careful proofreading my chapters made a substantial improvement in readability. As she is writing a book herself now, I am sure that, if she shows the same enthusiasm there as well, that book is going to be another piece of her work she (and I!) can be proud of.

I, Patrick Oonincx, would like to thank my loving fiancee Birgit Bobeck for running our family and having patience when I was spending weekends and holidays struggling with figures and formulas. Special thanks should go to her, because for the second time in our lives she is carrying a baby while I'm writing a book and, as in the previous case, both will see daylight around the same date. Last but not least

I have to make my apologies to my daughter Sacha for being a less playful dad in the last year. After this last sentence many formulas will be replaced by playtime.

Maarten Jansen, Eindhoven and Leuven, July 2004
Patrick Oonincx, Den Helder, July 2004

Contents

1. **The Classical Wavelet Transform for Continuous-time and Discrete-time Signals** 1
   1.1 The Continuous Wavelet Transform 1
       1.1.1 The Haar wavelet 3
       1.1.2 The Mexican Hat Wavelet 5
   1.2 The Discrete Wavelet Transform 7
       1.2.1 Discretizing the Continuous Wavelet Transform 7
       1.2.2 Multiresolution Analysis 8
       1.2.3 The Two-scale Equation 9
       1.2.4 Duality 12
       1.2.5 Vanishing Moments 13
   1.3 Multiresolution Analysis and Filterbanks 14
       1.3.1 The Decomposition Algorithm 14
       1.3.2 The Reconstruction Algorithm 16
       1.3.3 Implementation of the Discrete Wavelet Transform for Discrete-time Signals 17

2. **Second-generation Wavelets** 21
   2.1 Wavelets on Irregular Point Sets 21
   2.2 The Lifting Scheme 24
       2.2.1 The Haar Decomposition by Lifting 24
       2.2.2 The Lifting Scheme: Split, Predict and Update 26
       2.2.3 Basis Functions 29
       2.2.4 Subdivision 33
       2.2.5 Lifting Existing Wavelet Transforms 35
       2.2.6 Lifting and Polyphase 38
   2.3 The Construction of Second-generation Wavelets 40
       2.3.1 Multiscale Grids 41
       2.3.2 The Unbalanced Haar Transform 41
       2.3.3 Prediction Methods 45
       2.3.4 Updates for Vanishing Moments 52
   2.4 Lifting in Two (and More) Dimensions 53
       2.4.1 Definitions and Construction of Triangulations 54
       2.4.2 Delaunay Triangulation 54

		2.4.3	Multiscale Triangulations	55
		2.4.4	Multiscale Delaunay Triangulations	57

3. Nonlinear and Adaptive Lifting 61
- 3.1 Nonlinear Filters 61
  - 3.1.1 The Max-lifting Scheme 61
  - 3.1.2 The Median-lifting Scheme 64
- 3.2 Adaptive Lifting 66
- 3.3 Reconstruction of Adaptive Lifting 70
  - 3.3.1 Automatic Perfect Reconstruction 70

4. Numerical Condition 77
- 4.1 Stability in Wavelet Smoothing on Irregular Point Sets 78
- 4.2 Condition from Finite to Infinite Dimensions 81
  - 4.2.1 Condition Numbers 81
  - 4.2.2 Stable Bases 82
- 4.3 Numerical Condition of Wavelet Transforms 86
  - 4.3.1 General Stability Criteria 86
  - 4.3.2 Smoothness, Convergence (of Approximation) and Stability 89
- 4.4 Numerical Condition of Lifted Wavelet Transforms 91
  - 4.4.1 Lifting of Existing Stable Schemes 91
  - 4.4.2 Numerical Condition and Primal Vanishing Moments 92
  - 4.4.3 Stabilizing Updates 94
  - 4.4.4 Prediction 95
  - 4.4.5 Splitting Strategies 99

5. Applications of Nonlinear Lifting in Imaging 103
- 5.1 Image Retrieval Using Adaptive Lifting 103
  - 5.1.1 Quincunx Lifting 103
  - 5.1.2 Adaptive Lifting 106
  - 5.1.3 Redundant Lifting 108
  - 5.1.4 Feature Vectors 109
  - 5.1.5 Illustration of Image Retrieval 111
- 5.2 Adaptive Splitting using Normal Offsets 113
  - 5.2.1 $2D > (1D)^2$ 113
  - 5.2.2 Some Theoretical Background 118
  - 5.2.3 Normal Offsets 120

References 129

Index 135

1. The Classical Wavelet Transform for Continuous-time and Discrete-time Signals

Since its first appearance about 20 years ago the wavelet transform and its representations by means of filterbanks has been discussed in many books and papers, amongst which are some standard works [16, 22, 51, 58, 59, 73]. Although most readers of this book will be familiar with at least one of these publications, we will start with a wrap up of the wavelet transform as it has been used for the past two decades.

1.1 The Continuous Wavelet Transform

The continuous wavelet transform (CWT) transform was introduced in 1984 by Morlet and co-workers to analyse geophysical signals with some kind of modified windowed Fourier transform (WFT), which reads

$$\mathcal{F}_h[s](t,\omega) = \frac{1}{\sqrt{2\pi}} \int_{\mathbb{R}} s(y)\overline{h(y-t)}e^{-i\omega y}\,dy, \tag{1.1}$$

for $s \in L_2(\mathbb{R})$ a time-continuous signal and $h \in L_2(\mathbb{R})$ a window function. Here $L_2(\mathbb{R})$ denotes the space of square integrable functions on $\mathbb{R}$. The modification of the WFT was established by combining window function and Fourier mode $e^{-i\omega y}$ into one window function ψ that can be scaled. This modification became the CWT given by

$$\mathcal{W}_\psi[s](a,b) = \frac{1}{\sqrt{a}} \int_{\mathbb{R}} s(t)\overline{\psi\left(\frac{t-b}{a}\right)}\,dt, \tag{1.2}$$

for some wavelet function $\psi \in L_2(\mathbb{R})$, scaling parameter $a > 0$ and translation parameter $b \in \mathbb{R}$. Although this seemed to be a new transform, in the mathematical society this transform was already know as Calderón's reproducing formula [5].

The CWT can be written in a different way by introducing the translation operator $\mathcal{T}_b$ and dilation operator $\mathcal{D}_a$ on $L_2(\mathbb{R})$ by

$$\mathcal{T}_b[s](t) = s(t-b) \quad \text{and} \quad \mathcal{D}_a[s](t) = \frac{1}{\sqrt{a}}s\left(\frac{t}{a}\right), \tag{1.3}$$

for some $a \in \mathbb{R}^+, b \in \mathbb{R}$. Using these transforms we can also write (1.2) as

$$\mathcal{W}_\psi[s](a,b) = \langle s, \mathcal{T}_b \mathcal{D}_a \psi \rangle, \tag{1.4}$$

and by using Parseval's identity for the Fourier transform $\mathcal{F}$ also as

$$\mathcal{W}_\psi[s](a,b) = \langle \mathcal{F}s, \mathcal{F}\mathcal{T}_b \mathcal{D}_a \psi \rangle = \sqrt{a} \int_\mathbb{R} \hat{s}(\omega) \overline{\hat{\psi}(a\omega)}\, e^{ib\omega}\, d\omega, \tag{1.5}$$

with $\hat{s}$ the Fourier transform of the signal s.

¿From (1.4) two properties of the CWT can be obtained in a rather straightforward way, namely that the CWT is both bounded and continuous in its parameters (a,b). To show the continuity of the CWT we first assume ψ is a function of compact support. In this case ψ is uniformly continuous, meaning that

$$\|\mathcal{T}_b \mathcal{D}_a \psi - \mathcal{T}_{b'} \mathcal{D}_{a'} \psi\|_2 \to 0, \quad (a,b) \to (a',b'), \tag{1.6}$$

with $\|\cdot\|_2$ the L_2-norm. Since the space of all compactly supported functions in $L_2(\mathbb{R})$ is dense in $L_2(\mathbb{R})$, this result is inherited by all $\psi \in L_2(\mathbb{R})$, yielding the continuity of $\mathcal{W}_\psi[s]$ in its parameters (a,b) following (1.4). Furthermore, also from this definition we obtain by Schwarz's inequality

$$|\mathcal{W}_\psi[f](a,b)| \leq \|f\|_2 \cdot \|\psi\|_2 \; \forall_{a \in \mathbb{R}^+} \forall_{b \in \mathbb{R}},$$

i.e., the CWT is bounded for all (a,b).

Until now we have not discussed which functions can be used as an analyzing wavelet function ψ. One can think of functions that are compactly supported, like a real window that slides along the signals s. Mathematically speaking, a desirable property of a wavelet is that an inversion formula exists, i.e., s can be recovered from $\mathcal{W}_\psi[s]$. For deriving an inversion formula we start with a kind of Parseval's identity for the CWT. We integrate $|\mathcal{W}_\psi[s](a,b)|^2$ over $\mathbb{H} = \mathbb{R}^+ \times \mathbb{R}$ using the Haar measure $db\, da/a^2$. Furthermore, we write the CWT as a convolution (using the standard $u*v$ to denote the convolution of two functions u and v) $\mathcal{W}_\psi[s](a,b) = s * \mathcal{D}_a \check{\psi}(b)$, with $\check{\psi}(t) = \overline{\psi(-t)}$, which follows straightforward, from the definition of the CWT. Applying Parseval's identity on Fourier transforms, following [52], we arrive at

$$\begin{aligned}
\int_\mathbb{H} |\mathcal{W}_\psi[s](a,b)|^2\, db\, \frac{da}{a^2} &= \int_\mathbb{H} \left(|s * \mathcal{D}_a \check{\psi}(b)|^2\, db \right) \frac{da}{a^2} \\
&= 2\pi \int_\mathbb{H} |\hat{s}(\omega)|^2 |\hat{\psi}(a\omega)|^2\, d\omega\, \frac{da}{a} \\
&= \int_\mathbb{R} |\hat{s}(\omega)|^2 \left(2\pi \int_{\mathbb{R}^+} \frac{|\hat{\psi}(a\omega)|^2}{a}\, da \right) d\omega.
\end{aligned}$$

By defining

$$C_\psi(\omega) = 2\pi \int_{\mathbb{R}^+} \frac{|\hat{\psi}(a\omega)|^2}{a}\, da \tag{1.7}$$

we get the conservation of energy law for the CWT

$$\int_{\mathbb{H}} |\mathcal{W}_\psi[s](a,b)|^2 \, db \, \frac{da}{a^2} = C_\psi \|s\|_2^2, \qquad (1.8)$$

provided

$$0 < C_\psi < \infty, \qquad (1.9)$$

for almost all $\omega \in \mathbb{R}$. Condition (1.9) is called the admissibility condition and all $\psi \in L_2(\mathbb{R})$ that satisfy this condition are called (admissible) wavelets.

By polarization, (1.8) also yields

$$\begin{aligned}
\int_\mathbb{R} s(t)\overline{q(t)}\,dt &= \frac{1}{C_\psi} \int_\mathbb{H} \mathcal{W}_\psi[s](a,b)\overline{\mathcal{W}_\psi[q](a,b)}\,db\,\frac{da}{a^2} \\
&= \frac{1}{C_\psi} \int_\mathbb{H} \mathcal{W}_\psi[s](a,b) \int_\mathbb{R} 1/\sqrt{a}\,\overline{q(t)}\,\overline{\psi\left(\frac{t-b}{a}\right)}\,dt\,db\,\frac{da}{a^2}, \\
&= \int_\mathbb{R} \left(\frac{1}{C_\psi} \int_\mathbb{H} \mathcal{W}_\psi[s](a,b)\overline{\psi\left(\frac{t-b}{a}\right)}\,db\,\frac{da}{a^2\sqrt{a}} \right) \overline{q(t)}\,dt,
\end{aligned}$$

for all $q \in L_2(\mathbb{R})$. This results in the formal reconstruction formula

$$s(t) = 1/C_\psi \int_\mathbb{H} \mathcal{W}_\psi[s](a,b) \psi\left(\frac{t-b}{a}\right) db \, \frac{da}{a^2\sqrt{a}}, \qquad (1.10)$$

for all admissible wavelets ψ. Relation (1.10) holds weakly in $L_2(\mathbb{R})$. However, a stronger result also holds if both s and $\hat{s}$ are absolutely integrable as well. In that case it can be shown, e.g. see [52], that (1.10) also holds pointwise.

We conclude that inversion of the transform is only achieved for admissible wavelets ψ. The admissibility condition may seem hard to realize, however, every nonzero $\psi \in L_2(\mathbb{R})$, for which $\hat{\psi}$ is differentiable in 0 and $\hat{\psi}(0) = 0$, satisfies (1.9). Moreover, it can be shown that the set of admissible wavelets is dense in $L_2(\mathbb{R})$, e.g. see [56]. We discuss two well-known admissible wavelets that are used in many kind of applications.

1.1.1 The Haar wavelet

The first wavelet we discuss is the Haar wavelet, defined by

$$\psi(x) = \begin{cases} 1, & x \in [0, 1/2), \\ -1, & x \in [1/2, 1), \\ 0, & \text{otherwise.} \end{cases} \qquad (1.11)$$

In 1910, Haar already used this function for constructing an orthonormal basis in $L_2(\mathbb{R})$ by means of dilations and integer translations of a so-called *mother function*.

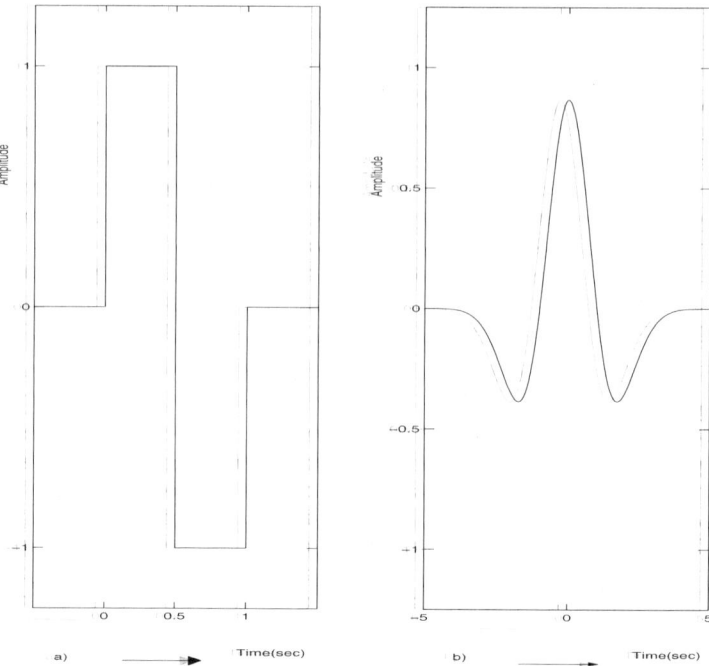

Figure 1.1. Two admissible wavelets: (a) the Haar wavelet, (b) the mexican hat wavelet.

Besides the orthonormality of a basis, Haar's concept coincides with the wavelet transform and its inversion formula. The aspect of basis functions is a topic that will be discussed in the next section. The Haar wavelet is depicted in Figure 1.1.(a).

Since for the Haar wavelet we have $\hat{\psi}$ differentiable in 0 and $\hat{\psi}(0) = 0$, the Haar wavelet indeed satisfies (1.9). However, we also show that the Haar wavelet is an admissible wavelet by computing

$$\hat{\psi}(\omega) = \frac{1}{\sqrt{2\pi}} \left(\int_0^{1/2} e^{-it\omega}\, dt - \int_{1/2}^1 e^{-it\omega}\, dt \right) = \frac{1}{\sqrt{2\pi}} \left(\frac{1 + e^{-i\omega} - 2e^{-i\omega/2}}{i\omega} \right),$$
(1.12)

and so

$$\frac{|\hat{\psi}(a\omega)|^2}{a} = \frac{|1 + e^{-ia\omega} - 2e^{-ia\omega/2}|^2}{a^3 \omega^2} = \frac{|e^{-ia\omega/2}|^2 \cdot |e^{ia\omega/4} - e^{-ia\omega/4}|^4}{a^3 \omega^2}$$

$$= 16 \frac{\sin^4(a\omega/4)}{a^3 \omega^2}.$$

Integrating by parts yields

1.1 The Continuous Wavelet Transform

$$\begin{aligned}
C_\psi &= \lim_{N \to \infty} \int_0^{N\omega/4} \frac{\sin^4(t)}{t^3}\,dt \\
&= \lim_{N \to \infty} \left. \frac{-\sin^4(t)}{2t^2} \right|_{t=0}^{N\omega/4} + \lim_{N \to \infty} 1/4 \int_0^{N\omega/4} \frac{2\sin(2t) - \sin(4t)}{t^2}\,dt \\
&= \lim_{N \to \infty} \left. \frac{\sin(4t) - 2\sin(2t)}{4t} \right|_{t=0}^{N\omega/4} + \\
&\quad \lim_{N \to \infty} \int_0^{N\omega/4} \frac{\cos(2t) - \cos(4t)}{t}\,dt \\
&= \lim_{N \to \infty} \left(\int_0^{N\omega/4} \frac{\cos(2t) - 1}{t}\,dt - \int_0^{N\omega/2} \frac{\cos(4t) - 1}{t}\,dt \right) \\
&= \lim_{N \to \infty} \operatorname{CI}(N\omega/4) - \ln(N\omega/4) - \operatorname{CI}(N\omega/2) + \ln(N\omega/2) = \ln 2,
\end{aligned}$$

where CI denotes the cosine integral see [76].

As an example we used the Haar wavelet to compute the CWT of a chirp-like signal $s(t) = \sin(\pi t^2)$ as depicted in Figure 1.2.(c). In Figure 1.2.(a) we depicted $|\mathcal{W}_\psi[s](a,b)|$ for $0 < a < 240$ and $0 < b < 8$ seconds. A very narrow Haar wavelet was chosen, so that even at the lowest scales, high resolution in time is obtained. This results, on the other hand, in a dispersion of the coefficients in scale, especially for the very first seconds of the signal. Furthermore, it is obvious that the energy in $\mathcal{W}_\psi[s]$ follows a convex hyperbola from high scale to low scale as time evolves, which is due to the fact that for this signal, frequency increases linearly in time and the fact that scale and frequency are reciprocal parameters. Finally, what can be noted in Figure 1.2.a is the appearance of aliasing at high scales and $4 < b < 8$. As the CWT can be written as a convolution product it can also be seen as a filter. In the case of the Haar wavelet we are dealing with a multiplication of the spectrum of s with a sinc function. This causes aliasing in the lower frequencies, i.e. the highest scales.

1.1.2 The Mexican Hat Wavelet

The second admissible wavelet we briefly discuss is the Mexican hat ψ, given by

$$\psi(t) = -\frac{d^2}{dt^2} e^{-t^2/2} = (1 - t^2)e^{-t^2/2}, \tag{1.13}$$

i.e., the second derivative of the Gaussian function, which also appears in the normal distribution in statistics. The Mexican hat is depicted in Figure 1.1.(b). We observe that although this function is not compactly supported it satisfies the admissibility

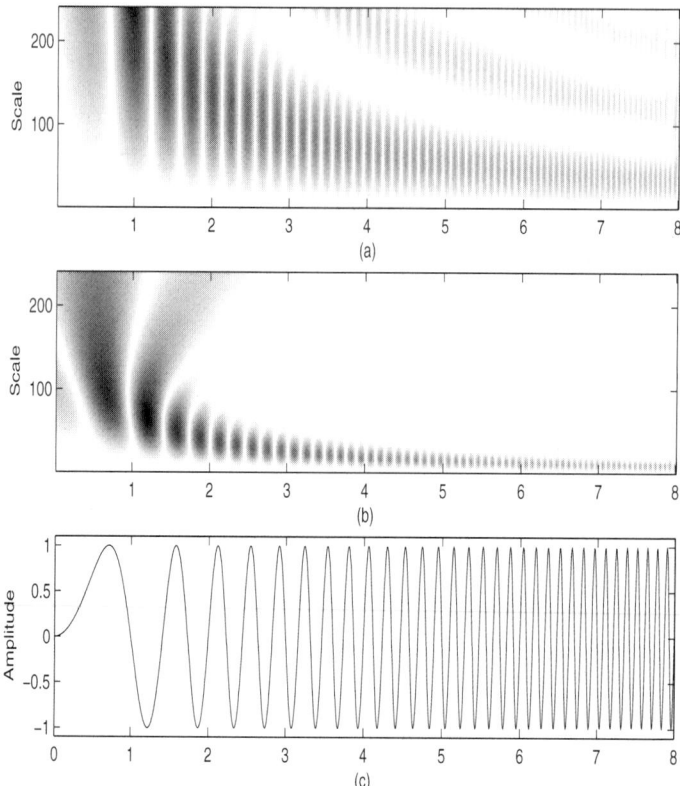

Figure 1.2. The CWT of a chirp signal using the Haar wavelet: (a) the modulus of the CWT $\mathcal{W}_\psi[s](a,b)$ with ψ the Haar wavelet, (b) as in (a) but now using the Mexican hat wavelet, (c) the chirp-like signal $s(t) = \sin(\pi t^2)$.

condition, which is shown by computing C_ψ. Since $\mathcal{F}\dfrac{d}{dt} = i\omega\mathcal{F}$ (where $\mathcal{F}$ stands for the Fourier transform) and

$$\int_R e^{-t^2/2} e^{-it\omega}\, dt = \sqrt{2\pi}\, e^{-\omega^2/2}, \tag{1.14}$$

we get $\hat{\psi}(\omega) = \dfrac{1}{\sqrt{2\pi}}\omega^2 e^{-\omega^2/2}$. Now a straightforward calculation yields

$$C_\psi = 2\pi \int_{R^+} \frac{|\hat{\psi}(a\omega)|^2}{a}\, da = \int_{R^+} a^3\omega^4 e^{-a^2\omega^2}\, da = 1/2 \int_{R^+} y e^{-y}\, dy = 1/2. \tag{1.15}$$

Also with the Mexican hat we have computed the CWT of the chirp-like signal $s(t) = \sin(\pi t^2)$. The modulus of the CWT is depicted in Figure 1.2.(b), for $0 < a <$

240 and $0 < b < 8$ seconds. A narrow mexican hat was chosen, resulting in a high resolution in time and dispersion of the coefficients in scale. The shape of the plot is similar to the plot above using the Haar wavelet, although two main differences can be observed. First, there is a larger spread in time, due to the fact that the Mexican hat has a larger width, i.e., it is not a function of compact support. Second, the aliasing problem disappeared, because in terms of Fourier transforms the spectrum $\hat{s}$ is multiplied with a rather narrow nearly finitely supported Gaussian-type function $\hat{\psi}$, which is obvious as the Gaussian function is an eigenfunction of the Fourier transform. In this context we would also like to recall that the Gaussian function gives best resolution in both time and frequency w.r.t. the Heisenberg inequality see [17].

1.2 The Discrete Wavelet Transform

1.2.1 Discretizing the Continuous Wavelet Transform

In the previous section, we considered the CWT. We showed that this integral transform is able to analyze signals both in time/space and scale. Moreover, it turned out that such signals can be recovered from their CWT if an admissible wavelet is used. In this section we show how to calculate efficiently the wavelet transform of a given signal s and to reconstruct it efficiently from its transform.

A first approach is to compute the wavelet transform only for a discrete subset $L \subset \mathbb{H}$, e.g.,

$$L = \{(a_0^j, kb_0 a_0^j) \mid j, k \in \mathbb{Z}\},$$

for some $a_0 > 1$ and $b_0 > 0$ and to replace the double integral in (1.10) by a double sum over L. Using such an approach we have to show that the integral representation is a redundant representation and that it can indeed be replaced by a double sum without loss of information. Furthermore, reconstruction of a signal $s \in L_2(\mathbb{R})$ by means of this double sum should depend continuously on s. This kind of stability is guaranteed if $\{\psi(a_0^j t - kb_0) \mid j, k \in \mathbb{Z}\}$ is a frame, i.e.,

$$m_F \|s\|_2^2 \leq \sum_{(a,b) \in L} |\mathcal{W}_\psi[s](a,b)|^2 \leq M_F \|s\|_2^2, \tag{1.16}$$

for some constants $m_F, M_F > 0$, independent of s, called the frame bounds. If we have a frame for which $m_F = M_F = 1$ and $\|\psi\|_2 = 1$ then the frame is an orthonormal basis in $L_2(\mathbb{R})$. In that case we can transform and reconstruct any signal s with respect to the lattice L. A common choice for L is a dyadic lattice using $a_0 = 2$ end $b_0 = 1$. In [21] Daubechies has given necessary and sufficient conditions on (ψ, a_0, b_0) for (1.16) to hold. Moreover, it was shown that ψ is an admissible wavelet for a CWT if (1.16) holds.

The transform using an orthonormal basis is a special case of the discrete wavelet transform (DWT). In general, the DWT is given by means on transforming a signal s using a basis

$$\{\,\psi(2^j t - k)\mid j,k\in\mathbb{Z}\,\}$$

and reconstructing it again to the original signal s. In particular, the decomposition and reconstruction of s should be established in a stable way. For ψ constituting a frame we have already seen that the wavelet coefficients continuously depend on s. By (1.16) the wavelet transform is computed from s in a stable way as $\mathcal{W}_\psi[s]$ depends continously on s. Reconstructing s from wavelet coefficients in a stable manner also asks for s to depend continously on its wavelet coefficients, i.e.,

$$m\|\boldsymbol{\alpha}\|_2^2 \leq \|s\|_2^2 \leq M\|\boldsymbol{\alpha}\|_2^2, \tag{1.17}$$

with α a double-indexed sequence of basis coefficients satisfying

$$s(t) = \sum_{j=-\infty}^{\infty}\sum_{k=-\infty}^{\infty} \alpha_{j,k}\,\psi_{j,k}(t), \tag{1.18}$$

where $\psi_{j,k}(t) = 2^{j/2}\psi(2^j t - k)$ and with constants $m, M > 0$. A basis of functions $\psi_{j,k}$ satisfying (1.17) is called a Riesz basis with Riesz constants m and M. A Riesz basis is sometimes also called an exact frame, since it can be shown that a Riesz basis satisfies the conditions of a frame and also contains "exactly enough" basis functions, instead of the redundancy that usually exists for a frame [62]. Furthermore, an orthonormal basis is a special case of a Riesz basis. Moreover, as for any type of basis, a Riesz basis can also be turned into an orthonormal basis by a Gramm Schmidt procedure. Therefore, it is sometimes also called semi-orthogonal basis.

1.2.2 Multiresolution Analysis

A way to construct a wavelet basis in $L_2(\mathbb{R})$ and to compute the basis coefficient of a signal s efficiently is given by the concept of a multiresolution analysis (MRA), due to Mallat [58] and Meyer [59]. It is a concept that was originally used as a signal-processing tool by means of perfect reconstruction filterbanks [73, 80]. The definition of such an MRA is given by an increasing sequence of closed subspaces $V_j, j \in \mathbb{Z}$, in $L_2(\mathbb{R})$,

$$\cdots \subset V_{-2} \subset V_{-1} \subset V_0 \subset V_1 \subset V_2 \cdots,$$

such that

1. $\bigcup_{j\in\mathbb{Z}} V_j$ is dense in $L_2(\mathbb{R})$,
2. $\bigcap_{j\in\mathbb{Z}} V_j = \{0\}$,
3. $f \in V_j \iff \mathcal{D}f = f(2\cdot) \in V_{j+1}, \ \forall j\in\mathbb{Z}$,
4. $f \in V_0 \iff \mathcal{T}f = f(\cdot - 1) \in V_0, \ \forall j\in\mathbb{Z}$,
5. $\exists \varphi \in L_2(\mathbb{R}) : \{T^k\varphi \mid k\in\mathbb{Z}\}$ is a Riesz basis for V_0,

with $\mathcal{D} := \mathcal{D}_{1/2}$ and $\mathcal{T} := \mathcal{T}_1$, following (1.3), and φ a real-valued function in $L_2(\mathbb{R})$, refered to as a scaling function.

Observe that the latter condition of an MRA equals the condition that there exists a scaling function φ such that $\{\mathcal{D}^j \mathcal{T}^k \varphi \mid k \in \mathbb{Z}\}$ is a Riesz basis for V_j, for any $j \in \mathbb{Z}$. This scaling function φ is often referred to as a *father function*. Obviously this follows directly from Condition 3 and from the fact that $\mathcal{D}$ is a unitary operator that does not effect the Riesz constants.

Constructing wavelet bases via an MRA is based on the inclusion $V_0 \subset V_1$. Obviously, we can define a subspace $W_0 \simeq V_1/V_0$. For a unique definition of W_0, we take W_0 perpendicular to V_0, giving $W_0 = V_1 \cap V_0^\perp$. Using the invariance of the subspaces V_j under the action of the unitary operator $\mathcal{D}$ we arrive in a natural way at the definition of the closed subspaces $W_j \subset L_2(\mathbb{R})$ by putting $W_j = V_{j+1} \cap V_j^\perp$. Recursively repeating the orthonormal decomposition of some V_J into V_{J-1} and W_{J-1} yields

$$\begin{aligned} V_J &= V_{J-1} \oplus W_{J-1} = V_{J-2} \oplus W_{J-2} \oplus W_{J-1} \\ &= \ldots = V_{-J} \oplus \left(\bigoplus_{j=-J}^{J-1} W_j \right). \end{aligned}$$

Taking $J \to \infty$ and applying Conditions 1 and 2 from the definition of an MRA yields

$$\bigoplus_{j \in \mathbb{Z}} W_j = L_2(\mathbb{R}).$$

This orthogonal decomposition of $L_2(\mathbb{R})$ is the key for constructing wavelet bases in $L_2(\mathbb{R})$, namely assume that a real-valued function $\psi \in V_1$ exists, such that $\{\mathcal{T}^k \psi \mid k \in \mathbb{Z}\}$ is a Riesz basis for W_0. This function is called a mother function, or wavelet function. Then $\{\mathcal{D}^j \mathcal{T}^k \psi \mid k \in \mathbb{Z}\}$ is a Riesz basis for W_j, for any $j \in \mathbb{Z}$. Owing to the mutual orthogonality of the subspaces W_j, we then have a Riesz basis in $L_2(\mathbb{R})$ given by the normalized functions $\{\mathcal{D}^j \mathcal{T}^k \psi \mid j, k \in \mathbb{Z}\}$. For ease of notation, we sometimes write $\psi_{j,k}$ for $\mathcal{D}^j \mathcal{T}^k \psi$ and similarly $\varphi_{j,k} = \mathcal{D}^j \mathcal{T}^k \varphi$.

¿From now on a function $\psi \in W_0$ that leads to this basis is refered to as a wavelet. Note that an admissible wavelet is not necessarily a wavelet that generates a Riesz basis by means of dilations and translations. For example, the Mexican hat wavelet is a widely used wavelet for computing the CWT of a signal s, but it is impossible to constitute a wavelet Riesz basis in $L_2(\mathbb{R})$ by means of dilating and translating one given Mexican hat wavelet. On the contrary, the Haar wavelet is a wavelet that is both admissible and constitutes a wavelet basis as constructed here. Moreover, the Haar wavelet yields an orthonormal basis in $L_2(\mathbb{R})$ by means of dyadic dilations and unit translates, which can be easily verified.

1.2.3 The Two-scale Equation

The previous construction of wavelet bases automatically leads to the question of finding a wavelet $\psi \in W_0$ that yields a Riesz basis by means of its unit translates.

We give an answer to this question by constructing such ψ based on the inclusions $V_0 \subset V_1$ and $W_0 \subset V_1$. Since both φ and ψ are assumed to be in V_0 and W_0 respectively, they are also both functions in V_1. Furthermore, we already observed that $\{\mathcal{DT}^k \varphi \mid k \in \mathbb{Z}\}$ is a basis for V_1, so that there exist sequences $\boldsymbol{g}$ and $\boldsymbol{h}$ such that

$$\varphi(t) = \sqrt{2} \sum_{k=-\infty}^{\infty} h_k \varphi(2t - k), \tag{1.19}$$

$$\psi(t) = \sqrt{2} \sum_{k=-\infty}^{\infty} g_k \varphi(2t - k). \tag{1.20}$$

Relations (1.19) and (1.20) are called two-scale equations or dilation relation and the sequences $\boldsymbol{g}$ and $\boldsymbol{h}$ are refered to as two-scale sequences or dilation sequences for ψ and φ respectively. Equation (1.20) is sometimes referred to as the wavelet equation, and equation (1.19) is then called the two-scale equation in a strict sense. Observe that for an MRA φ, and therefore $\boldsymbol{h}$, is known. Constructing a wavelet ψ therefore coincides with finding an appropriate sequence $\boldsymbol{g}$, as ψ then follows from its two-scale relation (1.20). An appropriate sequence $\boldsymbol{g}$ is a sequence related to ψ for which

- $\langle \varphi, T^k \psi \rangle = 0$ for all $k \in \mathbb{Z}$ ($V_0 \perp W_0$),
- $\{T^k \varphi \mid k \in \mathbb{Z}\} \cup \{T^k \psi \mid k \in \mathbb{Z}\}$ is a Riesz basis for V_1.

The second condition relates a Riesz basis in W_0 to the one in V_1, since this enables us to use the two-scale relations with sequences $\boldsymbol{g}$ and $\boldsymbol{h}$.

The first condition can be rewritten by using the two-scale relation for φ and ψ. We have

$$\begin{aligned} 0 = \langle \varphi, T^k \psi \rangle &= \sum_{l=-\infty}^{\infty} \sum_{n=-\infty}^{\infty} h_l \overline{g_n} \langle \varphi(2t - l), \varphi(2t - n - 2k) \rangle \\ &= \sum_{l=-\infty}^{\infty} \sum_{n=-\infty}^{\infty} h_l \overline{g_n} \tau_{2k+n-l} = \sum_{n=-\infty}^{\infty} (\boldsymbol{h} * \boldsymbol{\tau})_{2k+n} \overline{g_n} \\ &= \langle \boldsymbol{h} * \boldsymbol{\tau}, R^{2k} \boldsymbol{g} \rangle, \end{aligned} \tag{1.21}$$

with $\langle \cdot, \cdot \rangle$ the usual inner product on sequences, R the shift operator on sequences given by $Rg_n = g_{n-1}$ and $\boldsymbol{\tau}$ the sequence given by $\tau_n = \langle \varphi, T^n \varphi \rangle$. Note that in the case of an orthonormal basis generated by φ the first condition is translated into

$$\langle \boldsymbol{h}, R^2 \boldsymbol{g} \rangle = 0,$$

as $\tau_k = \delta_{k,0}$ for an orthonormal system.

The second condition can be translated into a condition on the two-scale sequences by means of the boundedly invertible operator S which maps signals in V_1 on their basis coefficients w.r.t. the Riesz basis $\{DT^k \varphi \mid k \in \mathbb{Z}\}$. By definition we have

1.2 The Discrete Wavelet Transform

$$Sf = \alpha \iff f(t) = \sum_{k=-\infty}^{\infty} \alpha_k \varphi(2t - k). \tag{1.22}$$

In particular, we have $S\varphi = h$ and $S\psi = g$. Furthermore, we observe that

$$\begin{aligned} Tf(t) &= \sum_{k=-\infty}^{\infty} \alpha_k \varphi(2t - k - 2) \\ &= \sum_{k=-\infty}^{\infty} \alpha_{k-2} \varphi(2t - k), \end{aligned}$$

and therefore $ST = R^2 S$. Since S is a boundedly invertible operator we immediately arrive at the following

$$S(\{T^k \varphi, T^k \psi \mid k \in \mathbb{Z}\}) \text{ is a Riesz basis for } S(V_1) = l_2(\mathbb{Z}), \tag{1.23}$$

where $l_2(\mathbb{Z})$ is the space of all square summable sequences. However, with the action of S on the wavelet and scaling functions, the latter can be rewritten as

$$\{R^{2k} h, R^{2k} g \mid k \in \mathbb{Z}\} \text{ is a Riesz basis for } l_2(\mathbb{Z}). \tag{1.24}$$

Together with (1.21) we now have two conditions on the sequences h and g. From these two conditions it is possible to find g once h is known.

To find a two-scale sequence g we transform both conditions by means of the z-transform, given by a discrete-time Fourier transform. Given a sequence g, then its z-transform reads

$$G(z) = \sum_{k=-\infty}^{\infty} g_k z^{-k}.$$

At this stage we will not elaborate further on the technique to rewrite (1.21) and (1.24) using this transform, but we only mention the final results as given in [62]. There, it was shown that the two-scale sequence g should satisfy two conditions, namely:

1. $T(z)G(z)\overline{H(z)} + T(-z)G(-z)\overline{H(-z)} = 0$ almost anywhere on the unit circle,
2. $\begin{pmatrix} H(z) & G(z) \\ H(-z) & G(-z) \end{pmatrix}$ is invertible for almost all z on the unit circle.

It can be verified in a straightforward way that possible choices for g are given in the z-domain by

$$G(z) = z^{2k+1} T(-z) \overline{H(-z)},$$

for any $k \in \mathbb{Z}$. In the case of an orthonormal basis we have $T(z) = 1$. Putting $k = 0$ yields $G(z) = z\overline{H(-z)}$, which results in $g_k = (-1)^k h_{1-k}$, also known as a pair of quadrature mirror filters. From now on we assume that both g and h, or equivalently ψ and φ, are known and we will concentrate on the issue of how to decompose and reconstruct signals in an efficient way using the framework of an MRA.

1.2.4 Duality

Before decomposing signals into wavelet bases we have to discuss one last topic, namely the issue of dual bases. This plays a role when not confining ourselves to an MRA based on orthonormal basis functions, as in such a case, we are not limited to a relatively small class of wavelets that can consitute such an MRA. However, orthonormality of the basis functions pays back when finding coefficients $s_{j,k}$ such that

$$s(t) = \sum_{j=-\infty}^{\infty} \sum_{k=-\infty}^{\infty} s_{j,k} \psi_{j,k}(t).$$

In case of orthonormality we straightforwardly have $s_{j,k} = \langle s, D^j T^k \psi \rangle$. In the non-orthogonal case we may look for a dual basis function $\widetilde{\psi}$, such that

$$\langle \psi, D^j T^k \widetilde{\psi} \rangle = \delta_{j,0} \delta_{k,0}$$

and $\{D^j T^k \widetilde{\psi} \mid j, k \in \mathbb{Z}\}$ is a Riesz basis for $L_2(\mathbb{R})$. For finding such dual wavelet bases we construct a dual MRA $(\widetilde{V}_j)$ with dual scaling function $\widetilde{\varphi}$. Such a dual MRA should satisfy the same conditions as the original one, with φ replaced by a dual scaling function $\widetilde{\varphi}$ satisfying

$$\langle \varphi, T^k \widetilde{\varphi} \rangle = \delta_{k,0}.$$

Furthermore, for the dual subspaces $\widetilde{V}_j$ and $\widetilde{W}_j$ the following relations should hold:

$$V_j \perp \widetilde{W}_j \quad \text{and} \quad \widetilde{V}_j \perp W_j,$$

corresponding to the biorthogonality of basis functions

$$\langle \varphi, T^k \widetilde{\psi} \rangle = 0 \quad \text{and} \quad \langle \widetilde{\varphi}, T^k \psi \rangle \quad \text{for all } k \in \mathbb{Z}. \tag{1.25}$$

This cross-perpendicularity also implies

$$W_i \perp \widetilde{W}_j, \quad i \neq j,$$

which is equivalent with the duality property

$$\langle \psi, D^j T^k \widetilde{\psi} \rangle = \delta_{j,0} \delta_{k,0}. \tag{1.26}$$

Since the dual MRA has equivalent properties as the original MRA there also exist two-scale relations relating scaled dual basis functions to an original dual scaling function. So there also exist sequences $\widetilde{g}$ and $\widetilde{h}$, such that

$$\widetilde{\varphi}(t) = \sqrt{2} \sum_{k=-\infty}^{\infty} \widetilde{h}_k \widetilde{\varphi}(2t - k), \tag{1.27}$$

$$\widetilde{\psi}(t) = \sqrt{2} \sum_{k=-\infty}^{\infty} \widetilde{g}_k \widetilde{\varphi}(2t - k). \tag{1.28}$$

We will not elaborate further on how to construct such biorthogonal basis functions, but we would like to refer to [13], in which the widely used Cohen-Daubechies-Fauveau (CDF) biorthogonal wavelet functions are introduced. This family of wavelet bases is constructed from B-spline scaling (father) functions. The wavelets themselves are splines (i.e., piecewise polynomials with specific smoothness properties in the knots) as well. Some members of the family, such as the CDF 2,2 and the CDF 4,4 with less dissimilar lengths (also known as the CDF 9,7) are particularly popular in image processing. The nomenclature within the family (e.g., CDF 2,2) mostly refers to the so-called vanishing moments, discussed in the next section. As another example of such biorthogonal basis functions, in Section 5.1.1 Neville interpolation functions are used for an application in image processing. In the remainder of this chapter, we assume that a dual MRA with corresponding dual basis functions is available.

1.2.5 Vanishing Moments

Several properties may be desired for a scaling function and its related mother wavelet, e.g. the existence of an orthonormal basis by means of integer translates is one of them. As we have seen in the previous section, orthogonality can be incorporated into the definition of an MRA by replacing the requirement of a Riesz basis. Then, dual basis functions are not needed for finding basis coefficients to develop a given signal into a series of wavelet basis functions.

Another very important property of a wavelet function is the existence of so-called vanishing moments. We say that a wavelet basis function $\psi_{j,k}$ has N vanishing moments if

$$\int_{-\infty}^{\infty} t^p \, \psi_{j,k}(t) \, dt = 0, \tag{1.29}$$

for $0 \leq p < N$. This property can also be translated into a certain kind of "blindness" of the basis function w.r.t. polynomials up to order N. If one has to analyze data consisting of random data superposed on a linear trend, then wavelet analysis will only consider the random data if the wavelets are blind to constant and linear functions, i.e., if the wavelet functions have two vanishing moments. The latter is the case for the Mexican hat wavelet that can be used for the CWT. In case of the Haar wavelet, only constant functions are vanished by the analysis, which means in practice that the mean of a given signal will not be taken into account in the wavelet analysis.

Up to now we have only mentioned vanishing moments of the wavelet $\psi_{j,k}$. However, if such a wavelet does not constitute an orthonormal basis, an interesting property may be a high number of vanishing moments for the dual basis functions $\widetilde{\psi}_{j,k}$. Because the series expansion of a given signal s by means of non-orthogonal basis functions $\psi_{j,k}$ reads

$$s(t) = \sum_{j=-\infty}^{\infty} \sum_{k=-\infty}^{\infty} \langle s, \widetilde{\psi}_{j,k} \rangle \psi_{j,k}(t),$$

vanishing moments of the dual basis functions cause basis coefficients $\langle s, \widetilde{\psi}_{j,k}\rangle$ to vanish for certain polynomial types of signal s. Polynomials or signals that "look like" polynomials are well approximated by a truncated (or thresholded) wavelet expansion if the dual basis functions have a sufficiently high number of vanishing moments. Vanishing moments related to the dual MRA will be refered to as dual vanishing moments, and vanishing moments of the original basis functions will be called primal vanishing moments. In Chapter 4 the importance of primal vanishing moments for numerical reasons will be discussed.

1.3 Multiresolution Analysis and Filterbanks

An MRA can be related to *filterbanks* by regarding a one-level decomposition and reconstruction $V_{j+1} = V_j \oplus W_j$. At the same time, this relation yields a scheme to calculate the wavelet transform of a function in $L_2(\mathbb{R})$ and to reconstruct it from its transform in a fast way. For this decomposition and reconstruction we introduce the orthoprojectors $\mathcal{P}_j$ and $\mathcal{Q}_j$ on V_j and W_j respectively. By definition we have

$$\mathcal{P}_{j+1} = \mathcal{P}_j + \mathcal{Q}_j, \qquad (1.30)$$

for all $j \in \mathbb{Z}$. Note that Conditions 1 and 2 from the definition of an MRA yield

$$\lim_{j \to \infty} \mathcal{P}_j s = s \qquad (1.31)$$

and

$$\lim_{j \to -\infty} \mathcal{P}_j s = 0 \qquad (1.32)$$

for all $s \in L_2(\mathbb{R})$.

1.3.1 The Decomposition Algorithm

For deriving a decomposition scheme we consider one single decomposition step. We assume $\mathcal{P}_{j+1} s \in V_{j+1}$ is known for a given signal $s \in L_2(\mathbb{R})$ and for a certain $j \in \mathbb{Z}$. Consequently, there exists a sequence s_{j+1} such that

$$\mathcal{P}_{j+1}s = \sum_{k=-\infty}^{\infty} s_{j+1,k} \varphi_{j+1,k}. \qquad (1.33)$$

Moreover, the sequence s_{j+1} is given by $s_{j+1,k} = \langle \mathcal{D}^{j+1} \mathcal{T}^k \widetilde{\varphi}, s \rangle$, using the dual scaling function $\widetilde{\varphi}$. Following decomposition (1.30) we have

$$\mathcal{P}_{j+1}s = \sum_{k=-\infty}^{\infty} s_{j,k} \varphi_{j,k} + \sum_{k=-\infty}^{\infty} d_{j,k} \psi_{j,k}, \qquad (1.34)$$

with the sequence d_j given by

1.3 Multiresolution Analysis and Filterbanks

$$\mathcal{Q}_j s = \sum_{k \in \mathbb{Z}} d_{j,k} \psi_{j,k}.$$

¿From the two-scale equation (1.27) we have

$$\varphi_{j,k} = \sum_{n=\infty}^{\infty} h_n \varphi_{j+1,2k+n}, \tag{1.35}$$

and a similar expression obviously holds for the dual two-scale equation. Note that moving from $\varphi_{j,0}$ to $\varphi_{j,k}$ at scale j corresponds to a move from $\varphi_{j+1,n}$ to $\varphi_{j+1,2k+n}$. Using the known sequence $s_{j+1,k}$ we may derive using (1.33) and the dual version of (1.35) that

$$\begin{aligned} s_{j,k} &= \langle P_{j+1}s, \widetilde{\varphi}_{j,k} \rangle = \sum_{n=-\infty}^{\infty} \widetilde{h}_n \langle P_{j+1}s, \widetilde{\varphi}_{j+1,2k+n} \rangle \\ &= \sum_{m,n=-\infty}^{\infty} s_{j+1,m} \widetilde{h}_n \langle \varphi_{j+1,m}, \widetilde{\varphi}_{j+1,2k+n} \rangle \\ &= \sum_{n=-\infty}^{\infty} s_{j+1,2k+n} \widetilde{h}_n = ((\downarrow 2)[s_{j+1} * \widetilde{\widetilde{h}}])_k, \end{aligned} \tag{1.36}$$

with $\widetilde{\widetilde{h}}_n = \widetilde{h}_{-n}$ and $(s_{j+1} * \widetilde{ph})_k = \sum_{n \in \mathbb{Z}} s_{j+1,k-n} \widetilde{\widetilde{h}}_n$, and with $(\downarrow 2)$ the downsampling operator given by

$$((\downarrow 2)[u])_k = u_{2k},$$

for all sequences u. In the same manner we get

$$d_j = (\downarrow 2)(s_{j+1} * \widetilde{\widetilde{g}}). \tag{1.37}$$

So s_j and d_j are obtained from s_{j+1} by taking s_{j+1} as an input sequence for discrete-time filters given with filter coefficients $\widetilde{\widetilde{h}}$ and $\widetilde{\widetilde{g}}$ respectively. After filtering, the output sequences are downsampled by a factor 2. The decomposition algorithm, as described in this manner, can be visualized by means of the analysis part of a two-channel filterbank as depicted in Figure 1.3.

Expressions for s_{j-n} and d_{j-n} for $n \geq 1$ can also be obtained recursively, namely

$$s_{j-n} = ((\downarrow 2)\mathcal{C}_{\widetilde{h}})^n s_j \quad \text{and} \quad d_{j-n} = (\downarrow 2)\mathcal{C}_{\widetilde{g}}((\downarrow 2)\mathcal{C}_{\widetilde{h}})^{n-1} s_j, \tag{1.38}$$

with $\mathcal{C}_u$ denoting filtering with coefficients u. So s_{j-n} and d_{j-n} can be easily obtained by iterative use of the analysis side of the two-channel filterbank as depicted in Figure 1.3.

The implementation of this algorithm to obtain wavelet coefficients d_j is known in literature as the fast wavelet transform (FWT), which is of computational complexity $\mathcal{O}(N)$. As a comparison, the widely used fast Fourier transform (FFT) already uses a number of computations of order $N \log N$.

16   1. The Classical Wavelet Transform for Continuous-time and Discrete-time Signals

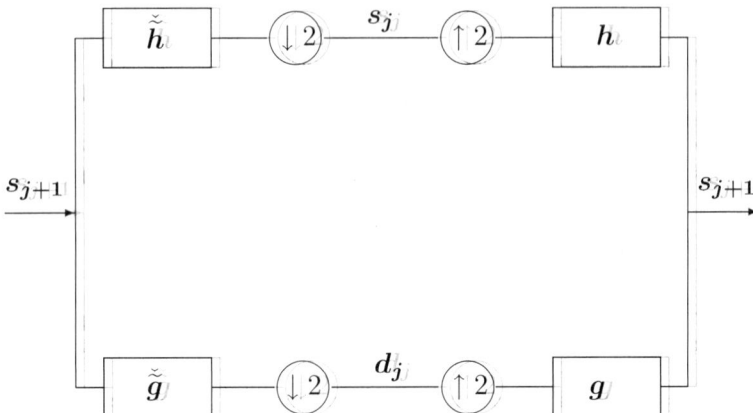

Figure 1.3. A two-channel filterbank related to the wavelet filters $h\tilde{h}$, g and $\tilde{g}$. Data s_{j+1} are decomposed into detailed data d_j^i and approximation data s_j by means of dual filters in the analysis part of the filterbank. Reconstruction of the original data is achieved in the synthesis part of the filterbank (right).

1.3.2 The Reconstruction Algorithm

Once the decomposition of $\mathcal{P}_{j+1}s$ into $\mathcal{P}_j s$ and $\mathcal{Q}_j s$ is established by means of s_j and d_j^i, it is also interesting to have an efficient algorithm to recover s_{j+1} out of s_j and d_j^i. To derive such an efficient method for reconstruction we first represent $\mathcal{P}_j s$ and $\mathcal{Q}_j s$ in terms of $\varphi_{j+1,k}$, $k \in \mathbb{Z}$, the basis functions of V_{j+1}. For this we introduce for $s \in L_2(\mathbb{R})$ the l_2-sequences

$$\alpha_{n,m,k} = \langle \mathcal{P}_n s, \tilde{\varphi}_{m,k} \rangle \quad \text{and} \quad \beta_{n,m,k} = \langle \mathcal{Q}_n s, \tilde{\varphi}_{m,k} \rangle \tag{1.39}$$

Since $\mathcal{P}_j s$ is also a signal in V_{j+1}, we can write

$$\mathcal{P}_j s = \sum_{k=-\infty}^{\infty} \alpha_{j,j+1,k} \varphi_{j+1,k}.$$

The sequence $\alpha_{j,j+1,k}$ can be rewritten, using (1.34) and (1.35), as

$$\begin{aligned}
\alpha_{j,j+1,k} &= \langle \mathcal{P}_j s, \tilde{\varphi}_{j+1,k} \rangle = \sum_{n=-\infty}^{\infty} s_{j,k} \langle \varphi_{j,n}, \tilde{\varphi}_{j+1,k} \rangle \\
&= \sum_{m,n=-\infty}^{\infty} s_{j,n} h_m \langle \varphi_{j+1,2n+m}, \tilde{\varphi}_{j+1,k} \rangle \\
&= \sum_{n=-\infty}^{\infty} c_{j,n} h_{k-2n} = (((\uparrow 2)c_j) * h)_k,
\end{aligned} \tag{1.40}$$

with $(\uparrow 2)$ the upsampling operator given by

$$(\uparrow 2)[\boldsymbol{u}](k) = \begin{cases} u_{k/2}, & k \bmod 2 = 0, \\ 0, & \text{otherwise}, \end{cases}$$

for all sequences $\boldsymbol{u}$. We observe that $(\downarrow 2)(\uparrow 2) = \mathcal{I}$: downsampling an upsampled signal is the identity. The reverse, upsampling a downsampled sequence, is not the identity.

Similar obersations for $\mathcal{Q}_j s$ yield

$$\mathcal{Q}_j s = \sum_{k=-\infty}^{\infty} \beta_{j,j+1,k} \varphi_{j+1,k},$$

with

$$\beta_{j,j+1}^j = ((\uparrow 2) d_j^j) * g, \tag{1.41}$$

Resuming, we obviously derived the reconstruction formula

$$s_{j+1} = \alpha_{j,j+1} + \beta_{j,j+1}^j = ((\uparrow 2) s_j) * h + ((\uparrow 2) d_j^j) * g.$$

Hence, s_{j+1} can be recovered from $\alpha_{j,j+1}$ and $\beta_{j,j+1}$. Analoguous to the decomposition side, we can relate the reconstruction algorithm to a two-channel filterbank. For this we take $\alpha_{j,j+1}$ and $\beta_{j,j+1}$ as output sequences of two filters given by their filter coefficients h and g respectively and with input sequences s_j and d_j upsampled by a factor 2. This can be visualized by means of the synthesis part of a two-channel filterbank, as has been depicted also in Figure 1.3.

Recursively, we get expressions for $\alpha_{j-n,j}$ and $\beta_{j-n,j}$ for $n \geq 1$, namely

$$\alpha_{j-n,j} = (\mathcal{C}_h(\uparrow 2))^n s_{j-n} \text{ and } \beta_{j-n,j} = (\mathcal{C}_h(\uparrow 2))^{n+1} \mathcal{C}_g(\uparrow 2) d_{j-n}. \tag{1.42}$$

In terms of filterbanks we can say that $\alpha_{j-n,j}$ and $\beta_{j-n,j}$ are obtained by iterative use of the synthesis part of the two-channel filterbank of Figure 1.3. The recursive approach to obtain $\alpha_{j-n,j}$ and $\beta_{j-n,j}$ is depicted in Figure 1.4 for $n = 2$.

The complete algorithm of decomposing and reconstructing functions by means of filterbanks related to an MRA is also known in literature as the pyramid algorithm.

1.3.3 Implementation of the Discrete Wavelet Transform for Discrete-time Signals

A problem that appears in decomposing a function $s \in L_2(\mathbb{R})$ at several scales by means of the pyramid algorithm is that the coefficients s_j should be known in order to compute $\mathcal{P}_{j-m} s$, $m \in \mathbb{N}$, by means of the pyramid algorithm. Computing s_j not only slows down the algorithm, it may also turn out to be impossible if only samples s of a signal s_c are known. In such cases, at a fixed resolution level j, one can identify the coefficients s_j with the sampled time-continuous signal s_c at resolution 2^j, i.e.,

18   1. The Classical Wavelet Transform for Continuous-time and Discrete-time Signals

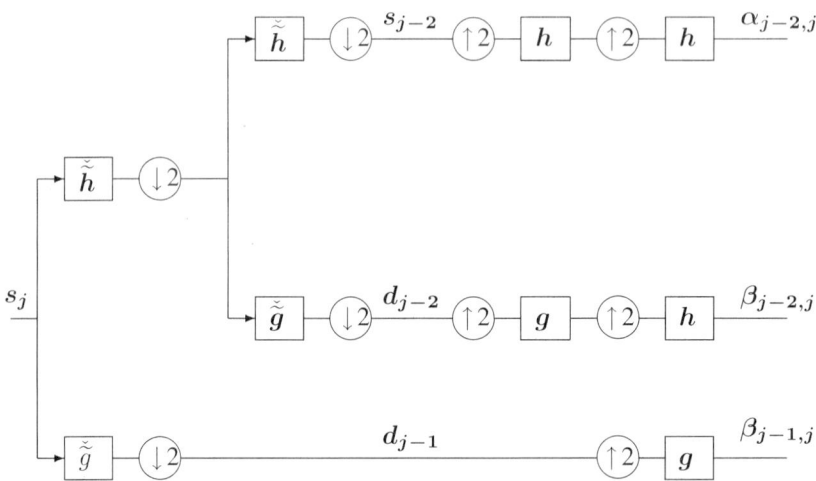

Figure 1.4. Decomposition and reconstruction by means of a filterbank at two resolution levels. The analysis part of an MRA-related filterbank is used on approximation data at the first level to obtain approximation and detail data at a second level (s_{j-2}, d_{j-2}). For the synthesis part, the same is done without merging intermediate results.

$$s_{j,k} = s_c(2^{-j}k).$$

We show that under certain conditions on the time-continuous signal s_c and the scaling function φ the above is a good approximation for the coefficients s_j.

We first assume that $s_c \in L_2(\mathbb{R})$ is Hölder continuous of order $\alpha \in (0,1]$, i.e.,

$$|s_c(t_0) - s_c(t_1)| \leq C \cdot |t_0 - t_1|^\alpha, \tag{1.43}$$

for all $t_0, t_1 \in \mathbb{R}$ and for a constant $C > 0$. (For $\alpha > 1$, the definition has to be extended to a form equivalent to the Definition 5.1 of Lipschitz continuity.) ¿From the Hölder continuity of s_c and Jackson's inequality, e.g. see [81], it follows that

$$\|\mathcal{P}_j s_c - s_c\|_\infty \leq C \sup_{0 < |h| < 2^{-j}} \|s_c - \mathcal{T}_h s_c\|_\infty$$
$$\leq C_0 2^{-\alpha j} \to 0 \ (j \to \infty),$$

for some positive constants C and C_0 and $\|\cdot\|_\infty$ the supremum norm. In particular we have $|\mathcal{P}_j s_c(t) - s_c(t)| \to 0$ for any value of t.

Next, we construct an operator $\mathcal{S}_j$ similar to $\mathcal{P}_j$, but now using the sampled signal as coefficients instead. This reads

$$\mathcal{S}_j[s_c](t) = \sum_{k \in \mathbb{Z}} s_c(2^{-j}k)\varphi(2^j t - k).$$

Similar to $\mathcal{P}_j$ we intend to give an estimate of difference between $\mathcal{S}_j s_c$ and the sampled signal s. To do this we need additional constraints on the scaling function φ, namely

$$\sum_{k \in \mathbb{Z}} \varphi(t-k) = 1 \text{ for almost any } t \in \mathbb{R}, \tag{1.44}$$

known as the partition of unity, and

$$\sum_{k \in \mathbb{Z}} |k|^\alpha |\varphi(k)| < \infty. \tag{1.45}$$

These two conditions on φ which may seem quite strong constraints. However, Condition (1.45) is already satisfied if φ is compactly supported, which is the case for the well-known Daubechies functions [22] and the biorthogonal spline scaling functions see [8]. For φ continuous in $\mathbb{Z}$, Condition (1.44) is satisfied in the case of an orthonormal MRA with $\hat{\phi}(0) = 1/\sqrt{2\pi}$ and ϕ absolutely integrable, e.g. see [56]. More on this partition of unity (in a second-generation context) follows in Section 4.4.2.

Given these two conditions we derive

$$\begin{aligned}
|\mathcal{S}_j[s_c](2^{-j}n) - s_c(2^{-j}n)| &= \left| \sum_{k \in \mathbb{Z}} s_c(2^{-j}k)\varphi(n-k) - s_c(2^{-j}n) \right| \\
&= \left| \sum_{k \in \mathbb{Z}} (s_c(2^{-j}k) - s_c(2^{-j}n))\varphi(n-k) \right| \\
&\leq \sum_{k \in \mathbb{Z}} |s_c(2^{-j}n) - s_c(2^{-j}(n-k))| \, |\varphi(k)| \\
&\leq C 2^{-j\alpha} \sum_{k \in \mathbb{Z}} |k|^\alpha |\varphi(k)| \to 0 \; (j \to \infty).
\end{aligned}$$

Combining the two results obtained, yields

$$\begin{aligned}
|\mathcal{P}_j[s_c](t) - \mathcal{S}_j[s_c](t)| &= |\mathcal{P}_j[s_c](t) - s_c(t) - \mathcal{S}_j[s_c](t) + s_c(t)| \\
&\leq |\mathcal{P}_j[s_c](t) - s_c(t)| + |\mathcal{S}_j[s_c](t) - s_c(t)| \to 0,
\end{aligned}$$

for $j \to \infty$. This result guarantees that replacing scaling coefficients with observations does not harm the convergence of the approximation for Hölder continuous functions with $\alpha \leq 1$. On the other hand, further analysis [73] shows that, if we have Lipschitz functions with a higher degree of smoothness, i.e., if $\alpha > 1$, a wavelet approximation shows faster convergence if the (dual) wavelet basis has sufficient vanishing moments. This faster convergence is then slowed down, back to the level of a simple Haar analysis, if scaling coefficients are replaced with observations. The resulting approximation is still a convergent one, but the benefits from the vanishing moments are lost. Strang and Nguyen [73] call this the *wavelet crime*. This issue shows up in a few examples of the subsequent Chapter.

2. Second-generation Wavelets

This chapter discusses the construction of so-called second-generation wavelets using the *lifting scheme*. We first illustrate the need for non-classical wavelets by an example in the field of noisy data smoothing: classical wavelets are not the appropriate tool when the data are observed on irregular grids. Next, in Section 2.2, we develop the lifting methodology. The actual construction of second-generation wavelets on irregularly observed (sampled) data follows in Section 2.3. In Section 2.4, special attention is paid to additional problems in the construction of second-generation wavelets in two dimensions.

The lifting scheme can be introduced both in a typical signal processing way, i.e., using diagrams, and in a formal, mathematical way, i.e., using operators, or at least matrices. This chapter presents both points of view. We are aware that readers mostly prefer one way or the other, but we also believe that both approaches offer specific insight into the mechanism behind lifting. And although lifting might look easy to understand from the pictures (at least to readers familiar with thinking in pictures and diagrams), full insight into basis functions might go far beyond that first view.

2.1 Wavelets on Irregular Point Sets

We start this chapter with an illustration from data smoothing. Suppose we are given a sequence of observations y_i of a function $f(x)$. The observations are subject to additive noise and they do not lie on an equidistant grid:

$$y_i = f(x_i) + \varepsilon_i, \qquad i = 1, \ldots, N.$$

The number of observations, N, is not necessarily dyadic (i.e., an integer power of 2). Since classical wavelets are constructed on regular grids, we could act as if the y_i are observed on equidistant points. Suppose for a moment that $N = 2^J$ is dyadic. Figure 2.1(a) illustrates what happens. The figure depicts a detail of an estimation of the underlying curve along with the non-equidistant observations. The locations x_i of the observations were taken from a uniform density on the interval. The true underlying curve is the well-known "heavisine" test function, $f(x) = 4\sin(4\pi x) - \text{sign}(x - 0.3) - \text{sign}(0.72 - x)$ [39] and the estimation was obtained by a very classical wavelet thresholding procedure [38]: wavelet coefficients (apart from those at the three coarsest scales) below a given threshold are replaced by zero. The threshold value taken in this example is the universal threshold $\lambda = \sqrt{2 \log N} \sigma$, where σ is the standard deviation of the noise in ε_i. This threshold value has been designed to guarantee (asymptotically and almost surely) a *smooth* output. The choice of the wavelet functions used in this example has also been inspired by the objective of smoothness: it is the Coiflet [23, Section 4, page 511] with four vanishing moments. The details of the properties of this particular wavelet are not so important for this discussion (they can be found in the literature); it is sufficient here just to note that this function is smooth. As Section 2.2.3 discusses in the framework of lifting wavelet transforms, it holds that smooth wavelet functions lead to smooth reconstructions.

Altogether, we may not have the optimal algorithm with respect to error compared with the true underlying function, but we did pay a lot of attention to the smoothness of the output. Yet, as we see on the figure, the result is non-smooth, even in locations away from the singularity (where we could expect some wiggling effects, Gibbs phenomena). If we start from an equal number N of equidistant observations, however, the result is much smoother, as illustrated in Figure 2.1(b).

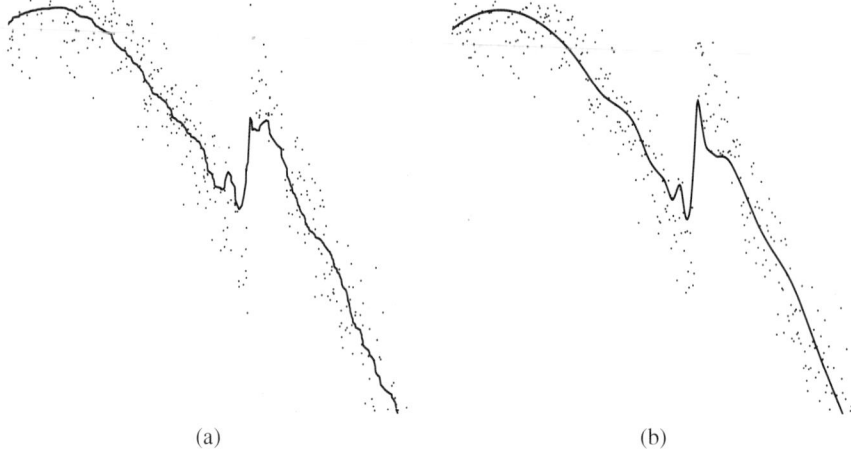

(a)   (b)

Figure 2.1. (a) Smoothing non-equispaced data as if they were equispaced yields a non-smooth output: the structure of the grid is reflected in the output, since it was not taken into account during the procedure. (b) Smoothing equispaced data does not have to deal with this problem.

This map-to and remap-from a regular grid can be described as follows: suppose that the grid locations x_i are function values of an increasing function $x(u)$ that maps a regular grid $u_i = i/N$ onto the non-equispaced points $x_i = x(u_i)$. The wavelet procedure then *actually* operates on the function $g(u) = f(x(u))$; it constructs an approximation $\tilde{g}(u)$ for $g(u)$ as

$$\tilde{g}(u) = \sum_k s_{J,k} \, \varphi_{J,k}(u),$$

where $s_{J,k}$ are the scaling coefficients at the finest scale and $\varphi_{J,k}$ are the corresponding scaling basis functions. After remapping, using the inverse function $u(x)$, this yields an approximation

$$\tilde{f}(x) = \sum_k s_{J,k}\, \varphi_{J,k}(u(x)).$$

Since the basis functions $\varphi_{J,k}(u)$ have been designed to be smooth with respect to their argument, the functions $\tilde{\varphi}_{J,k}(x) = \varphi_{J,k}(u(x))$ are not smooth: they reflect the structure of the grid, which shows up in Figure 2.1(a). This observation follows a general rule of thumb when using wavelet (or any other basis) expansions: use basis functions that look like the underlying function that you approximate. If we expect (or know) f to be (piecewise) smooth, it is not a good idea to use non-smooth basis functions. Also note that this simple mapping-remapping is generally impossible in two (or more) dimensions where scattered data (i.e., tw-dimensional (2-d) irregularly spaced observations) cannot be associated in a unique and consistent way to points at regular locations.

It is possible to *construct* the basis functions on regular grids, while *evaluating* them in the actual, irregular points x_i [2]:

$$\tilde{f}(x_i) = \sum_k s_{J,k}\, \varphi_{J,k}(x_i).$$

This, however, corresponds to an additional pre- or post-processing of the data, where one cannot rely on the fast algorithms for discrete wavelet transforms. This pre- and/or post-processing may also undo interesting statistical properties of the data, such as independency of the observations, thereby leading to the use of more general and possibly suboptimal data-processing routines. This pre- and post-processing can be seen as an interpolation of the original data into a regular grid. Such an interpolation can also be carried out explicitly [30, 44, 53]. In statistical literature, the statistical distribution of the grid locations X_i (now random variables, and therefore, denoted in capital letters) is sometimes explicitly taken into account. Such a model is referred to as *random design*. A model where the x_i are not considered as random variables is then referred to as *(irregular but) fixed design*. When the randomness of X_i is modelled explicitly, it is possible to reason about the expected function values in the *expected* locations EX_i, and the corresponding wavelet and scaling coefficients. This reasoning is used in a statistical motivation for using simple mapping and remapping when the grid locations X_i can be modelled as instances from a uniform distribution on the interval, as was the case in Figure 2.1 [4]. Although the result may be satisfactory in terms of expected values, it clearly requires an additional smoothing in practical applications.

In this chapter, we introduce a new type of wavelet decomposition, so-called second-generation wavelets, that are constructed and evaluated directly on irregular locations. As a bonus, we get a procedure that does not need a dyadic (power of 2) number of input values. The approach also has the possibility to automatically adapt to data on finite intervals without relying on artificial techniques, like symmetric or periodic extension of the data near the boundaries. The philosophy behind these second-generation wavelets is the so-called *lifting scheme*.

2.2 The Lifting Scheme

There are several ways to introduce the concept of lifting with two basic points of view: the first one introduces lifting as a method to *construct* (or re-implement) a wavelet transform, such as the Haar transform, whereas the second concentrates on the ability of lifting to *enhance* (hence the name *lifting*) an existing wavelet transform by adding desirable properties. This section starts with a description from the first point of view and with the simple case of a lifted version for the Haar transform. This is a fairly intuitive approach. A more formal approach, stated in terms of matrices, follows in Section 2.2.5. Readers with a preference for formal, mathematical treatments can skip the intermediate sections.

2.2.1 The Haar Decomposition by Lifting

Like every wavelet transform, a Haar decomposition takes *fine-scale scaling coefficients* as input. For the moment, we can think about these scaling coefficients as the observations. The reason why we call these input coefficients *fine-scale* coefficients has to be seen in the context of what a wavelet analysis is going to do with it: it is going to destilate the global — in other words, coarse scale — trends out of it. Since such an analysis cannot create a more detailed view than the resolution level on which the observations come in, the resolution level of the input is the finest level. That level is denoted by an integer $J \in \mathbb{Z}$. The choice of J is, in principle, arbitrary (the other levels of the analysis will be assigned a lower number), but it is mostly chosen such that the lowest possible level of the analysis corresponds to zero. In classical wavelet literature, the notion of scale is sometimes defined as $s = 2^J$. This comes from the continuous wavelet transform theory, where scale is a continuous notion and the resolution levels in a discrete wavelet transform are a (dyadic) discretization of it. These chapters mostly use the notions of scale and resolution level as synonyms.

A Haar decomposition — just as all classical wavelet transforms — proceeds in successive steps, where each step further corresponds to one (coarser) resolution level. Each step of a Haar decomposition computes averages and differences of adjacent input values, i.e., let $s_{j+1,k}$ be the input at scale $j+1$, then one step of a Haar decomposition transforms this into averages $s_{j,k}$ and details (differences) $d_{j,k}$ at scale j:

$$s_{j,k} = \frac{s_{j+1,2k} + s_{j+1,2k+1}}{2} \quad (2.1)$$

$$d_{j,k} = s_{j+1,2k+1} - s_{j+1,2k}. \quad (2.2)$$

The next step of the decomposition then repeats the same transform on the averages $s_{j,k}$. Those averages are a coarser resolution version of the input, in other words, i.e., in signal processing terminology, it is a low-pass filtered version. The details carry the information that was present in the fine-scale version but get lost by averaging: they allow one to reconstruct the original fine-scale version by the inverse transform. The first of each pair of fine-scale input values is the average minus half of the difference, whereas the second fine-scale value can be found as the average plus half of the difference:

$$s_{j+1,2k+1} = s_{j,k} + d_{j,k}/2 \qquad (2.3)$$
$$s_{j+1,2k} = s_{j,k} - d_{j,k}/2 \qquad (2.4)$$

From this reconstruction, it immediately follows that the average $s_{j,k}$ can be written as the first value plus half of the difference between the second and first values, i.e.:

$$s_{j,k} = s_{j+1,2k} + d_{j,k}/2. \qquad (2.5)$$

This observation leads to a new version of the algorithm. Instead of computing averages and differences *simultanuously*, we *first* compute the differences and *next* the averages. This approach can be visualized in the diagram in Figure 2.2. The diagram shows a transform in three stages. First the data is split into even and odd indexed observations. Next, the differences between adjacent odds and evens are computed. These differences are stored instead of the odds. Indeed, using expression (2.5), the odds are no longer necessary to find the averages. Those averages are computed in the final step according to (2.5), and can then replace the evens. The diagram thus visualises the following operations:

$$\begin{aligned} \text{difference} &= \text{odd} - \text{even} \\ \text{average} &= \text{even} + \frac{\text{difference}}{2}. \end{aligned}$$

This is the *lifting scheme* implementation of the Haar transform. It replaces the classical filterbank scheme of Figure 1.3. For reasons that become clear further in this chapter, the calculation of the detail (difference) coefficients is an example of what is called a *dual* lifting step, while the average calculation is a typical example of a *primal* lifting step.

This reorganization of the Haar transform has a fundamental impact on the *interpretation* of the coefficients. Indeed, in the classical implementation, the differences are computed *together with* the averages, and they are interpreted as the information you need to reconstruct the input *from the averages*. In the lifting scheme implementation, averages and differences are computed *separately*; we could easily omit the computation of the averages and instead proceed with the evens themselves (although it is not a good idea to do so, as we discuss later). So, *before* we compute the averages, the differences stand for the information we need to reconstruct the complete input, this time not from the averages (simply because we have no averages yet), but from the *evens*. This might look like a purely philosophical discussion at this point, but further on we discuss the importance of this argument in understanding the mechanism behind lifting. This discussion is also the key to the extension of

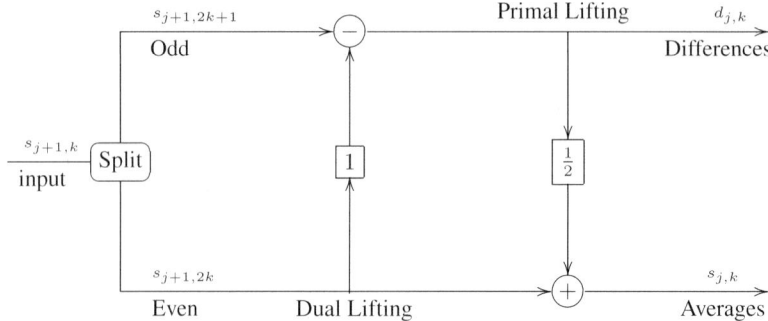

Figure 2.2. The lifting scheme implementation of the Haar transform. This scheme is equivalent to the classical filterbank of Figure 1.3. The rectangular boxes linking the odd and even branches stand for *convolution* (i.e., filter) operations: in the dual lifting step, the even indexed coefficients are convolved with some sequence and the result is subtracted from the odd coefficients. In the particular case of a Haar transform, that convolution sequence is a Kronecker sequence, represented by the figure 1 in the diagram. As a result, the dual lifting step simply subtracts (unmodified) evens from odds. The primal lifting step of a Haar decomposition convolves the resulting differences with one-half times a Kronecker sequence and adds the result to the even coefficients. Convolutions can also be seen as matrix-vector multiplications, where the matrix has a Toeplitz structure. The matrices used in dual and primal steps of a Haar transform are the identity matrix and one-half times the identity matrix respectively.

the lifting approach beyond Haar transforms. The differences are viewed as details, or offsets of the odds from the evens. They indicate how far the observed odd value deviates from what you predict by looking at the evens only. This *prediction* is quite simple here: every odd indexed observation is predicted by the value in its left even neighbour. Using another predictor, based on more even observations than just the immediate even neighbour on the left-hand side, is a first step towards a general lifting scheme.

2.2.2 The Lifting Scheme: Split, Predict and Update

A general lifting scheme consists of three types of operation: a split, followed by a (mostly alternating) sequence of *dual* and *primal* lifting operations. The splitting step partitions the observations into two disjoint sets. For now (but not forever), these sets correspond to observations with even and odd indices.

Next, as illustrated in Figure 2.3, the odd indexed input values are replaced by the offset (i.e., the difference) between those values and a prediction based on the evens only. The idea is that the observations show a significant redundancy, and a prediction of one half by the other half results in a *sparse representation*: a few large details indicate the locations of features, such as jumps or cusps, that could not be well predicted. The Haar transform predictor is a simple constant extrapolation from

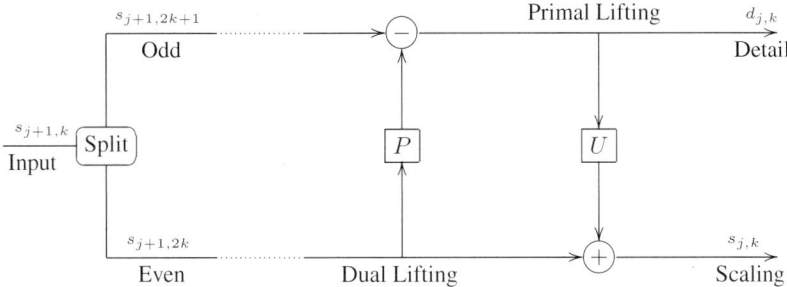

Figure 2.3. A general lifting scheme is a (mostly alternating) sequence of primal and dual lifting steps, initiated by a splitting stage. Note that general convolutions replace the identity operations in a Haar transform. The convolutions are represented by their matrices P and U. For a classical wavelet transform, these matrices are Toeplitz, but in the second-generation setting, the "convolution" becomes time dependent, so the matrices get an arbitrary structure. The dots between the splitting stage and the dual lifting step in the diagram indicate that a general lifting scheme may consist of more than one sequence of dual and primal lifting steps: after a primal step, a new dual step may follow. It is not necessary that primal and dual steps come in pairs: the total number of primal and dual steps may be unequal, although in most cases, primal and dual steps do alternate.

the left even neighbour. A slightly more sophisticated approach is to interpolate in even neighbours on both sides of an odd observation. Linear prediction only takes immediate neighbours; cubic interpolation, as in Figure 2.4, also involves second-order neighbours. The prediction is always based on even observations only. Intermediate odd observations are not used for predictions in other odd locations, since *all* odd observations are replaced by a detail coefficient. If this observation was used to predict another one, that prediction cannot be reconstructed upon synthesis of the original input. For reasons still to be explained in Section 2.3.3, this prediction step is known as *dual lifting*.

As illustrated in the previous section for the Haar transform, most multiscale decompositions then proceed with a so-called *primal lifting step*, or *update step*. As we discuss in Chapter 4, such an update step is absolutely necessary for a stable wavelet transform. If this step is missing, the even indexed values proceed unchanged to the next, coarser scale. If this happens scale after scale, a few input values completely determine the coarse-scale analysis. We would rather prefer a situation where the coarse-scale coefficients are based on averaged values of the input. This way, fluctuations on single observations do not get a large-scale impact. In signal processing terms, we want the coarse-scale coefficients to represent a low-pass filtered version of the input. Without additional averaging, even coefficients are simply downsampled input values, which causes aliasing (ringing). For now, the update of the evens can thus be seen as an anti-aliasing step.

The first primal lifting step may be followed by a new dual step, and so on. This is indicated in Figure 2.3 by the vertical dashed lines and the dotted connection

28   2. Second-generation Wavelets

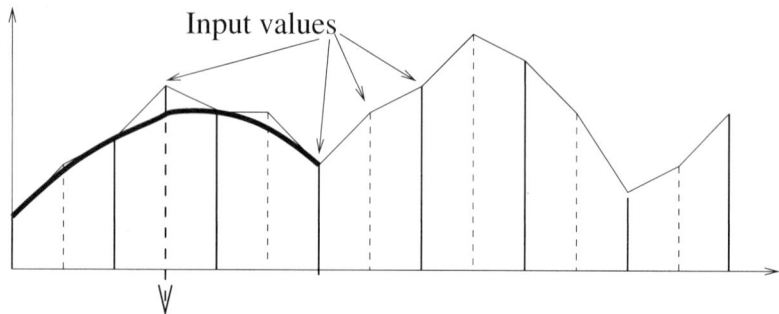

Prediction by cubic interpolation in this point

Figure 2.4. A cubic interpolation as a prediction operator. The thin, piecewise linear line links the input data. The bold line is an example of a cubic polynomial, interpolating four successive points with even index.

between the splitting and the first depicted dual lifting step. Each step adds more properties — for instance more sparsity, i.e., more (near-) zero coefficients — to the overall transform: it is a gradual increase in complexity; hence the name lifting.

The lifting scheme has a few immediate interesting properties:

1. It is *in-place*, meaning that in all stages; one of the input branches (the even or odd side) can be overwritten by the output of that step. There are no *simultaneous* calculations, as in the classical filterbank implementation. This property saves working memory in large data applications.
2. The structure of the lifting scheme makes it not only suitable for in-place calculations, but also readily invertible. The inverse lifting scheme is depicted in Figure 2.5. Its construction is based on the observation that every single lifting step can be easily undone. Indeed, when odd input values are replaced by their offsets (i.e., differences or details) from a prediction, we keep the even inputs on which this prediction is based. So, we can recompute that prediction and add it to the offsets to reconstruct the odd inputs. The same idea holds in the update step: since it preserves the detail coefficients used for the construction of the update, we can recompute the update from the data *after* this step as well as we could *before* the update step.

So, the forward transform

$$\begin{aligned} \text{detail} &= \text{odd} - P(\text{even}) \\ \text{scaling} &= \text{even} + U(\text{scaling}) \end{aligned}$$

is simply inverted by

$$\begin{aligned} \text{even} &= \text{scaling} - U(\text{scaling}) \\ \text{odd} &= \text{even} + P(\text{even}) \end{aligned}$$

This is not possible in the classical filterbank implementation; the filters on the input values are not readily invertible. The inverse filters are the solution of a linear system, and one trick to solve this system easily is to work in the Fourier domain; in the Fourier domain, filtering (i.e., convolution) becomes simple multiplication, which makes the solution of this linear system particularly easy.

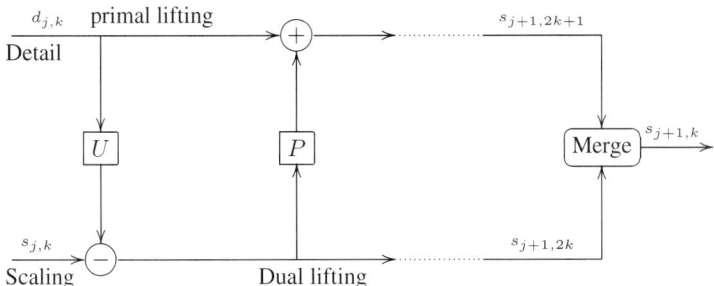

Figure 2.5. The inverse lifting scheme. Owing to the specific structure of the lifting scheme, this diagram can be obtained by reflecting the forward scheme along a vertical axis, and substitute plus with minus signs and vice versa.

2.2.3 Basis Functions

Until now we have concentrated on the lifting operations and their effect on the *values* of the resulting coefficients. In a multiresolution analysis, these values can also be *interpreted* as coefficients in a basis decomposition. More specifically, the inputs are viewed as fine-scale scaling coefficients in the decomposition

$$f_J(x) = \sum_{k \in \mathbb{Z}} s_{J,k} \varphi_{J,k}(x), \tag{2.6}$$

where $\varphi_{J,k}(x)$ are *scaling basis functions* at resolution level J. If the number of basis functions is infinite, then basis decompositions are also called (basis) *expansions*. A typical example is a Fourier expansion or Fourier *series*. Also, a full wavelet decomposition of a function is considered in an infinite dimensional function space (typically the space of square integrable functions), which means that you need an infinite number of wavelet basis functions with corresponding wavelet coefficients to represent a general function. Obviously, in practice, we have only a limited number of observations, so we can represent those discrete data with a finite number of coefficients; all coefficients at finer scales than the resolution of observation can be thought being zero.

These wavelet basis functions are fully determined by the adopted wavelet transform, in our case from the lifting operations. In other words, we cannot just *choose* an interpretation. How exactly the underlying basis functions follow from the lifting operations is the subject of this section. In this context, a wavelet transform itself is a change of basis. The function $f_J(x)$ is rewritten as

$$f_J(x) = \sum_{k \in \mathbb{Z}} s_{L,k} \varphi_{L,k}(x) + \sum_{j=L}^{J-1} \sum_{k \in \mathbb{Z}} d_{j,k} \psi_{j,k}. \tag{2.7}$$

Again, the wavelet basis functions $\psi_{j,k}$ follow from the transform. The properties of these basis functions determine how we can interpret the coefficients and, consequently, how we can manipulate them. For instance, as we discuss in Section 4.4.2, it is important for numerical reasons that $\psi_{j,k}$ has a vanishing integral, and for some applications we also like higher vanishing moments, i.e., we like

$$\int_{-\infty}^{\infty} x^p \psi_{j,k}(x) dx = 0.$$

If the integral of $\psi_{j,k}$ is non-zero, then manipulation of the coefficients (for instance for compression or noise reduction) may have unpredictable, undesirable effects after reconstruction.

If we want to design a wavelet analysis with given properties (such as vanishing integrals and moments), it is important to get insight into how the lifting scheme shapes the basis functions. This nontrivial question is the key to full understanding of the lifting mechanism. The key behind keeping track of the basis functions is the following trick. The decomposition (2.7) holds for $f_J(x) = \psi_{j,k}$, for fixed j and k, with all coefficients equal to zero, except $d_{j,k} = 1$. In other words, we can study the basis functions by investigating what a corresponding unit coefficient stands for. For instance, what happens with this unit coefficient if we run an inverse transform and express $\psi_{j,k}$ back in terms of fine scale scaling basis functions, as in Equation (2.6)?

In each step of a lifting transform we have two branches of coefficients. Every coefficient in every intermediate step not only has a *value* (which is obvious), but also an *interpretation*. We start again from the Haar example. For the moment we conjecture that the input f_k can be interpreted as coefficients $s_{J,k}$ in a basis with indicator functions, as shown in Figure 2.6 on top. The function $f_J(x) = \sum_k s_{J,k} \varphi_{J,k}$ is then a piecewise constant (i.e., a staircase) function, with function levels given by the observations f_k. As discussed below, this representation is consistent with the Haar decomposition, and in Section 2.2.4 we discuss a procedure, called subdivision, that allows one to verify that inidicator basis functions are indeed the correct interpretation for scaling coefficients in a Haar decomposition.

In Section 2.2.1 we already explained that the interpretation of a difference coefficient depends on the question whether it has to be related to an even value or rather an average. *Before* the update step, we have evens and differences, so the difference indicates how much the odd differs from the even. As illustrated in Figure 2.6, this can be expressed in terms of basis functions. We start with Haar scaling functions at fine scale. After the Haar prediction, it is as if we extend the even values over the adjacent odd intervals. In other words, we take the even values as a prediction for both even and odd intervals. The odd scaling functions are still there to correct for this extending on the odd intervals. Only after the update do we see that the detail expresses (up to a factor $1/2$) how much the odd value is below and (at the same time and with the same amount) how much the next, even value is above the common average. This below and above results in the Haar wavelet basis functions. Once again, the *interpretation* of a coefficient, its meaning or background, is nothing else but the corresponding basis function in the *synthesis* wavelet basis. That basis function describes the unit contribution of a coefficient to a given signal.

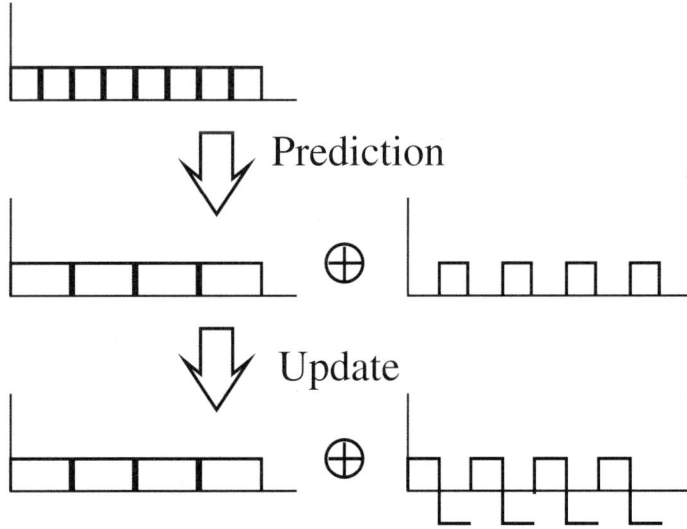

Figure 2.6. Intermediate basis functions in the lifted version of a Haar transform. After the prediction step, we have computed how far the odd samples are from the even ones. This corresponds to extending the even basis functions over the joint even and odd intervals. The wavelets, i.e., the detail basis functions, are just the odd scaling functions at the previous, fine level. These wavelets clearly have no vanishing moments. The update step changes the basis functions corresponding to the odd samples by updating the values of the even samples.

It is important to realise that, throughout the lifting scheme, the basis functions are generated in a different way and at different moments than the coefficient values. Indeed, after the Haar prediction step, we already have the detail values, but not yet the wavelet basis functions. The basis functions are simply the odd scaling functions at fine scale. Prediction preserves the even coefficient values and the odd basis functions.

To verify that this holds on a general lifting scheme, we concentrate on a scheme with one prediction step and one update step. The arguments are best explained on an inverse scheme, as in Figure 2.7. We want to study a wavelet $\psi_{j,k}$ and a scaling $\varphi_{j,k}$ basis function. This is, we want to know what a unit coefficient $d_{j,k}$ or $s_{j,k}$ stands for exactly. In the first instance, we want to express $\psi_{j,k}$ and $\varphi_{j,k}$ as a combination of scaling functions at finer scales. So, we run one step of the inverse transform with all zeros except one coefficient, i.e., $d_{j,l} = \delta_{l-k}$, and see what comes out. The scaling coefficients $s_{j+1,l} = g_{j,k,l}$ we get after one step are the coefficients in the wavelet equation:

$$\psi_{j,k} = \sum_l g_{j,k,l} \varphi_{j+1,l}.$$

This *wavelet* equation states that running the inverse transform from scale j with all coefficients zero, except $d_{j,k} = 1$, is equivalent to running the inverse transform starting from one scale finer, i.e., level $j+1$, but now with scaling coefficients $s_{j+1,l}$ equal to $g_{j,k,l}$. A similar argument leads to the *two-scale* equation for the scaling functions:

$$\varphi_{j,k} = \sum_l h_{j,k,l} \varphi_{j+1,l}. \tag{2.8}$$

If we are working on a regular grid (this is the classical filterbank situation), the coefficients $h_{j,k,l}$ are independent from scale and only dependent on the *relative* locations k and l, so (2.8) then reduces to the classical form:

$$\varphi_{j,k} = \sum_l h_{2k-l} \varphi_{j+1,l}.$$

Returning now to Figure 2.7, we know that in stage ①, we have the coarse-scale wavelet and scaling basis functions $\psi_{j,k}$ and $\varphi_{j,k}$. At stage ③, we have the fine-scale scaling functions $\varphi_{j+1,k}$. The question remains as to what the basis functions look like in between, i.e. at stage ②. We denote these intermediate basis functions by $\varphi_{j,k}^{[0]}$ and $\psi_{j,k}^{[0]}$. If we run the first step of this inverse lifting scheme, i.e., the update step, starting with all wavelet coefficients equal to zero, nothing changes. So, if one of the scaling coefficients is one, it still equals one after the update. In other words $\varphi_{j,k} = \varphi_{j,k}^{[0]}$. The update does not change the scaling functions used in the synthesis of $f(x)$. It does, however, have an impact on the wavelet basis functions $\psi_{j,k}$ in that synthesis. Indeed, if we run the (inverse) update step with $d_{j,l} = \delta_{l-k}$, we keep that Kronecker sequence on the detail branch, but also have coefficients, say $-\alpha_{j,k,l}, l \in \mathbb{Z}$, on the other branch. This is exactly the subset of update coefficients used to update several even values with the value of the detail coefficient $d_{j,k}$. So, running the inverse transform with a Kronecker sequence on the detail branch is equivalent to running the inverse *without update* (i.e., starting from stage ②), but then with additional scaling coefficients $-\alpha_{j,k,l}, l \in \mathbb{Z}$. Stated otherwise

$$\begin{aligned} \psi_{j,k} &= \psi_{j,k}^{[0]} - \sum_l \alpha_{j,k,l} \varphi_{j,k}^{[0]} \\ &= \varphi_{j+1,2k+1} - \sum_l \alpha_{j,k,l} \varphi_{j,k}. \end{aligned} \tag{2.9}$$

The update step changes the scaling coefficients, but also the wavelet *basis* $\psi_{j,k}$ used in the synthesis (reconstruction) of $f(x)$. This basis is also called the *primal* basis, as explained in Section 1.2.4, in contrast to the *dual* basis $\tilde{\psi}_{j,k}$, which consists of basis functions such that the *coefficients* of the decomposition can be found as

$$d_{j,k} = \langle f, \tilde{\psi}_{j,k} \rangle.$$

For this interpretation, the update is also called the *primal* lifting step.

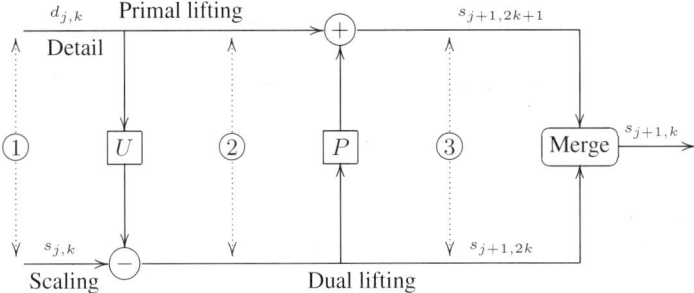

Figure 2.7. The evolution of the basis functions throughout the lifting scheme.

2.2.4 Subdivision

Section 2.2.3 has investigated how the basis functions within scale j are built up from finer scale scaling functions. This leaves us with the question of how these scaling functions look like. From the diagram in Figure 2.7, it is clear that if the scheme consists of one predictor and one update, then this update is of no importance for the scaling functions. If, however, the update is followed by another dual lifting step, all steps, dual and primal, play a role in the properties of the resulting scaling functions.

Those scaling basis functions can be found by running, scale by scale, the inverse transform with all detail coefficients zero, and starting from a Kronecker sequence as scaling coefficients at the level of interest. This is an iterative refinement process: every step adds new (odd indexed) locations and assigns a function value to them, based on values in the locations inserted in previous steps. Those existing values may change as well, at least if in the inverse scheme there exists a dual step preceding an update step. This is not the case in the scheme of Figure 2.7, where existing values proceed untouched to finer scales. In any case, if we continue this refining process up to infinity, then we find function values in all points of the grid (assuming for the moment that the iteration process converges). This refinement scheme is known as *subdivision*. It can be used to display any wavelet or scaling basis function, not only in the context of lifting, but also in the classical filterbank setting. Subdivision is also a well-known and studied technique for curve and surface generation in areas like computer aided geometric design (CAGD) and computer graphics. Examples of subdivision schemes are the *de Casteljau* algorithm for generating Bézier curves, or the *de Boor* algorithm for (more general) spline methods.

2. Second-generation Wavelets

Subdivision with interpolating prediction, such as illustrated in Figure 2.4, has been established as the Deslauries-Dubuc subdivision scheme [31, 32] *before* its application in wavelet decompositions, although it was soon recognized that this scheme could be interesting for the construction of wavelet transforms [40].

The result of subdivision with linear polynomial prediction is a piecewise linear spline, with knots in the data points at the scale where the subdivision is initiated. The basis functions can be found by running the scheme on a Kronecker sequence; they are first-order B-splines, as illustrated in Figure 2.8. If this linear prediction scheme is followed by the appropriate update step, then the resulting wavelet transform coincides with the popular CDF 2,2 spline wavelet decomposition. (Completing the statement in terms of issues to be discussed later on, that appropriate update step is a convolution with a sequence with two nonzero entries, where both of these degrees of freedom are used to have two primal vanishing moments.) This outcome of linear subdivision is the continuation of what we have seen with the lifting version of the Haar transform: the Haar scaling functions are indeed the zeroth-order B-splines, corresponding to a constant interpolation (actually an extrapolation) as predictor. Maybe a bit surprisingly, this line of B-spline results does *not* continue any further. The limiting function of a cubic subdivision scheme, for instance, is not a spline. Although any cubic polynomial lies within the space spanned by these basis functions, the basis functions themselves are not at all piecewise polynomial. An example of such a limiting function is depicted in Figure 2.9.

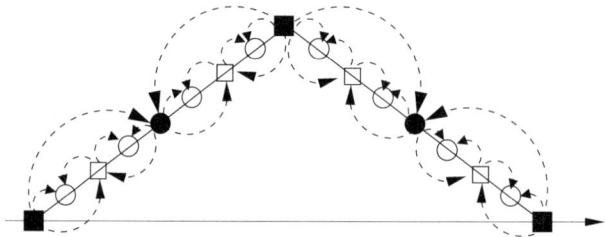

Figure 2.8. Linear subdivision leads to first-order B-spline basis functions. The initial grid is given by the filled squares. The initial data are a Kronecker sequence, in order to retrieve a basis function. The first step of the subdivision scheme is depicted with the filled circles. The next step corresponds to blank squares and the step thereafter yields the blank circles. The limit function is a piecewise linear basis function.

We emphasize once more that both the two-scale and wavelet equations on the one hand and the subdivision scheme on the other hand are entirely built on the *inverse* wavelet transform. The inverse transform is the key to retrieving the synthesis basis functions, i.e., the interpretation behind the coefficients in a multiscale decomposition. The two-scale and wavelet equations use one step of the inverse transform: they constitute the first iteration step in a subdivision scheme. It is important to realize that whenever one encounters a two-scale equation, this is just one step in a

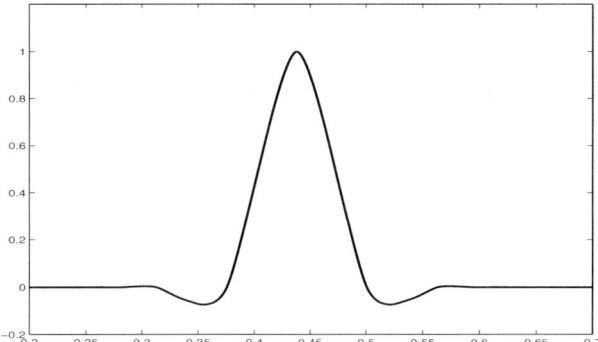

Figure 2.9. Limiting function of a cubic subdivision scheme (on a regular grid without boundaries). This function is not a piecewise polynomial, so it cannot be a spline function.

subdivision scheme, and subdivision is one — not the only [73] — method to solve it (at least numerically).

2.2.5 Lifting Existing Wavelet Transforms

Lifting can also be described in terms of matrices, rather than with diagrams. Suppose we are given a sequence of observations $f_i = f(x_i)$. For the moment, we start from the standard assumptions for a classical wavelet transform, namely, the grid locations x_i are equidistant and the number N of observations is dyadic, i.e., an integer power of two: $N = 2^J$. It will soon become clear that these restrictions are by no means necessary in a lifting setup.

Every step in a wavelet transform can be represented by a block-partitioned matrix

$$\tilde{W}_j = \begin{bmatrix} \begin{bmatrix} \tilde{H}_j \\ \hline \tilde{G}_j \end{bmatrix} & 0_j \\ \hline 0_j^T & I_j \end{bmatrix}.$$

If $\boldsymbol{w}_j$ is the vector of wavelet coefficients $w_{j,k}$ at scale j, and if $\boldsymbol{s}_j$ is the vector of scaling coefficients $s_{j,k}$ at scale j, then it transforms the vector

$$\begin{bmatrix} \boldsymbol{s}_{j+1} \\ \boldsymbol{w}_{j+1} \\ \boldsymbol{w}_{j+2} \\ \vdots \\ \boldsymbol{w}_{J-1} \end{bmatrix}$$

into the vector

$$\begin{bmatrix} s_j \\ w_j \\ w_{j+1} \\ w_{j+2} \\ \vdots \\ w_{J-1} \end{bmatrix}.$$

The matrices $\tilde{H}_j$ and $\tilde{G}_j$ represent the actual decomposition operations at this scale:

$$\begin{aligned} w_j &= \tilde{G}_j \cdot s_{j+1} \\ s_j &= \tilde{H}_j \cdot s_{j+1}. \end{aligned}$$

In the classical filterbank setting, these operations are convolutions, followed by downsampling (subsampling). Rows $2^{j+1} + 1$ to N are left untouched in this step: they contain wavelet coefficients at previous, finer scales. The full decomposition is then

$$\tilde{W} = \tilde{W}_0 \cdot \tilde{W}_1 \cdot \ldots \cdot \tilde{W}_{J-1}.$$

The inverse of the matrix $\tilde{W}_j$ has a similar block-partitioned structure:

$$W_j = \tilde{W}_j^{-1} = \left[\begin{array}{c|c} [\, H_j^T \mid G_j^T \,] & 0_j \\ \hline 0_j^T & I_j \end{array} \right].$$

From this, we can find that

$$s_{j+1} = H_j^T \cdot s_j + G_j^T \cdot w_j, \tag{2.10}$$

which corresponds to the classical filterbank operation. Since $W_j = \tilde{W}_j^{-1}$, it follows immediately that

$$H_j^T \cdot \tilde{H}_j + G_j^T \cdot \tilde{G}_j = I. \tag{2.11}$$

In order to retrieve the two-scale and wavelet equations for the synthesis or primal basis functions in this scheme, we define a function $f(x)$ as a synthesis from coefficients in that primal basis:

$$f(x) = \sum_{k \in \mathbb{Z}} s_{j+1,k} \varphi_{j+1,k}(x),$$

for some arbitrary coefficients $s_{j+1,k}$. This can be rewritten as an inner product:

$$f(x) = \Phi_{j+1} \cdot s_{j+1},$$

where $\Phi_{j+1} = [\ldots \varphi_{j+1,k}(x) \ldots]$ is a row vector of basis functions. We impose that this function can be rewritten as a combination of scaling functions at a coarser scale *plus* detail contributions:

$$f(x) = \Phi_j \cdot s_j + \Psi_j \cdot w_j.$$

2.2 The Lifting Scheme

Plugging in the reconstruction (2.10) into the equality

$$\Phi_{j+1} \cdot s_{j+1} = \Phi_j \cdot s_j + \Psi_j \cdot w_j,$$

yields the equality

$$\Phi_{j+1} \cdot H_j^T \cdot s_j + \Phi_{j+1} \cdot G_j^T \cdot w_j = \Phi_j \cdot s_j + \Psi_j \cdot w_j.$$

Since the coefficients $s_{j+1,k}$ were arbitrary, so are w_j and s_j. As a consequence, it must be that

$$\Phi_j = \Phi_{j+1} \cdot H_j^T, \quad (2.12)$$
$$\Psi_j = \Phi_{j+1} \cdot G_j^T. \quad (2.13)$$

This can be read — after taking the transposes of the matrices on the left- and right-hand side — as follows: the column vector of coarse-scale basis functions in the synthesis of a function can be found by applying a *forward* wavelet transform with the *synthesis* (or reconstruction, or inverse) filter coefficients.

From this set of matrices $H_j, G_j, \tilde{H}_j, \tilde{G}_j$, the lifting scheme constructs a new set $\widehat{H}_j, \widehat{G}_j, \widehat{\tilde{H}}_j, \widehat{\tilde{G}}_j$. There are two possible lifting operations:

1. <u>Primal lifting</u> defines the new matrices as

$$\widehat{H}_j = H_j$$
$$\widehat{\tilde{H}}_j = \tilde{H}_j + U_j \cdot \tilde{G}_j$$
$$\widehat{G}_j = G_j - U_j^T \cdot H_j$$
$$\widehat{\tilde{G}}_j = \tilde{G}_j,$$

for some *update* matrix U_j. Plugging in these definitions into the expressions for the coefficients reveals that

$$\widehat{s}_j = \widehat{\tilde{H}}_j \cdot s_{j+1} = (\tilde{H}_j + U_j \cdot \tilde{G}_j) \cdot s_{j+1} = s_j + U_j \cdot w_j$$
$$\widehat{w}_j = w_j,$$

and the new primal basis functions, used in the synthesis or reconstruction step, satisfy

$$\widehat{\Psi}_j = \Phi_{j+1} \cdot \widehat{G}_j^T = \Phi_{j+1} \cdot (G_j - U_j^T \cdot H_j)^T = \Psi_j - U_j \cdot \Phi_j$$
$$\widehat{\Phi}_j = \Phi_j.$$

It is an almost trivial exercise to verify [74, Theorem 8.1] that this new set of filters satisfies the perfect reconstruction property (2.11). This lifting operation can be represented by a diagram as in Figure 2.10.

2. <u>Dual lifting</u> defines the new matrices as

$$\begin{aligned} \widehat{H}_j &= H_j + P_j^T \cdot G_j \\ \widehat{\tilde{H}}_j &= \tilde{H}_j \\ \widehat{G}_j &= G_j \\ \widehat{\tilde{G}}_j &= \tilde{G}_j - P_j \cdot \tilde{H}_j, \end{aligned}$$

for some *prediction* matrix P_j. The impact on coefficients and basis functions can be found in a way similar to the discussion for primal lifting.

Primal lifting changes the primal wavelets, i.e., the basis functions used in the synthesis of a function, dual lifting changes the primal detail *coefficients*.

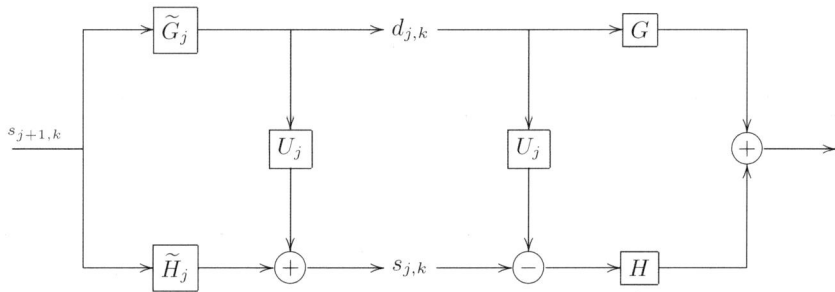

Figure 2.10. A primal lifting operation on an existing wavelet decomposition. The subsampling and upsampling step of the original filterbank is incorporated into the matrices $H_j, G_j, \tilde{H}_j, \tilde{G}_j$.

A lifted wavelet decomposition can be lifted again. Mostly, primal and dual lifting steps are added in an alternating chain. If the initial matrices H_j and $\tilde{H}_j$ are the even rows of the identity matrix and the matrices G_j and $\tilde{G}_j$ are the odd rows of the identity matrix, then the entire filterbank is constructed through lifting. The initial, trivial decomposition is known as the *lazy wavelet transform*, since it does nothing but split input into even and odd indexed observations.

2.2.6 Lifting and Polyphase

The classical filterbank scheme of a wavelet decomposition consists of convolutions followed by downsampling (subsampling) operations. Obviously, if it is implemented literally that way, then the subsampling throws away half of the work done in the convolution step. It is possible to skip the calculation of every other coefficient from the beginning, or, alternatively, we can reorganize the calculations

2.2 The Lifting Scheme

into smaller convolutions without subsampling. Indeed, let $(\downarrow 2)$ be the subsampling operator and let S denote the shift operator, i.e.,

$$y = Sx \Leftrightarrow y_k = x_{k+1},$$

and call s^e and s^o the even and odd subsignals of a signal s, i.e.,

$$s^e = (\downarrow 2)s_j \quad \text{of} \quad s^e_{j,l} = s_{j,2l}$$
$$s^o = (\downarrow 2)Ss_j \quad \text{of} \quad s^o_{j,l} = s_{j,2l+1}.$$

The classical filterbank implementation corresponds to

$$s_{j,k} = \sum_{l \in \mathbb{Z}} \tilde{h}_{l-2k} s_{j+1,l} = \sum_{l \in \mathbb{Z}} \hat{h}_{2k-l} s_{j+1,l}$$

$$w_{j,k} = \sum_{l \in \mathbb{Z}} \tilde{g}_{l-2k} s_{j+1,l} = \sum_{l \in \mathbb{Z}} \hat{g}_{2k-l} s_{j+1,l},$$

where, for ease of notation, we introduced the mirrored two-scale and wavelet equation coefficients $\hat{h}$ and $\hat{g}$ defined by $\hat{h}_k = \tilde{h}_{-k}$ and $\hat{g}_k = \tilde{g}_{-k}$. Using $a * b$ as a notation for the convolution of two sequences a and b, the classical filterbank analysis can also be written as

$$s_j = (\downarrow 2)(\hat{h} * s_{j+1})$$
$$w_j = (\downarrow 2)(\hat{g} * s_{j+1}).$$

It is a straightforward manipulation to rewrite this forward transform step as (the exercise might be easier by writing it up with full summations):

$$s_j = \hat{h}^e * s^e_{j+1} + S^{-1}\hat{h}^o * s^o_{j+1}$$
$$w_j = \hat{g}^e * s^e_{j+1} + S^{-1}\hat{g}^o * s^o_{j+1}.$$

This operation can be summarized in the diagram of Figure 2.11. The subsampling at the end has been substituted by a *splitting* stage in front. This *polyphase* implementation provides the link between the filterbank version and the lifting scheme. It can be proven that the polyphase operations can be further reorganized into a sequence of primal and dual lifting steps [28]. The procedure is based on a *Euclidean algorithm* for the greatest common divisor. That algorithm can be applied not only to numbers, but also to polynomials or power series. The coefficients of the power series in our case are given by the two-scale and wavelet equations. The intermediate relationships between remainder r, quotient q, dividend a and divisor b, namely

$$a = r + q \times b,$$

have the form of a lifting step: r is updated with q times the divisor to get a; alternatively, a is predicted by q times the divisor and r is the remainder, i.e., the offset, detail, or difference. The intermediate relationships yield the consecutive lifting steps.

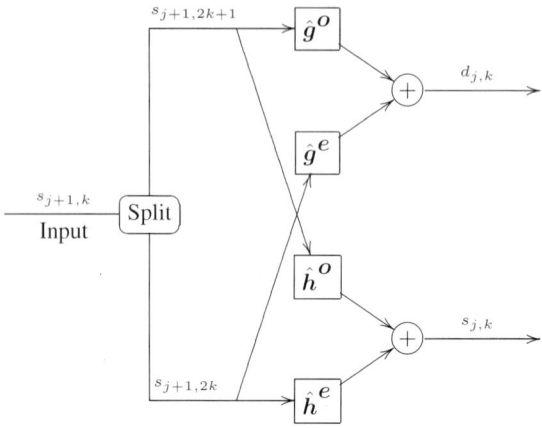

Figure 2.11. The polyphase implementation of a classical filterbank. The subsampling steps at the end have been replaced by a splitting stage in front. The polyphase implementation can be seen as an intermediate stage between the classical filterbank implementation and a full lifting version of a wavelet transform.

2.3 The Construction of Second-generation Wavelets

This section discusses the usage of lifting for the construction of wavelets on data with non-standard structures. In particular, we are interested in wavelets for observations on irregular time points (non-equispaced data, 2-d scattered data), wavelets for data on intervals (adapted to the interval boundaries), and wavelets for non-dyadic data (i.e., where the number of observations is not an integer power of two). The construction of such decompositions, based on lifting, is known as *second-generation wavelets*. Lifting also permits one to go even further and consider nonlinear and data-adaptive multiresolution decompositions. Since the transform becomes nonlinear, we can no longer describe these decompositions with wavelet *basis functions*. A well-known example here is the integer wavelet transform. After the introduction of the concept of multilevel grids, the discussion starts with an exploration of the Haar transform on irregular point sets. It turns out that the Haar decomposition can be easily generalized to this irregular setting, and it does not a priori involve lifting. Next, we proceed to the actual construction of second-generation wavelets. Elements of this construction are the dual lifting steps, discussed in Section 2.3.3, primal lifting steps, discussed in Section 2.3.4 and the order in which those steps appear. The guidelines for the construction of lifting steps given in this chapter are by no means complete. For instance, Chapter 4 discusses how to construct second-generation wavelet decompositions that are numerically stable. In general, it holds that the application at hand plays an important role in the design of a suitable sequence of lifting steps. The guidelines presented in this chapter are general methods and ideas in the design of such a transform.

2.3 The Construction of Second-generation Wavelets

2.3.1 Multiscale Grids

Each step in the classical wavelet decomposition is based on an equal partition of the input grid into "even" and "odd" locations. This happens explicitly in the splitting stage of a lifting scheme and a bit less explicitly in the subsampling stage of a classical filterbank implementation. The lifting implementation does not impose an equal partition. It is not always possible, anyway, if we allow a non-dyadic number of observations and do not rely on techniques like symmetric extension to deal with this. Also, the irregular structure of the grid may call for a nonequal partition. This is explored further in Section 4.4.5. Throughout this text, we keep talking about "evens" and "odds", although this may refer to a more general partition.

A general multilevel (multiscale) grid is defined as a sequence of strictly increasing sets X_j of locations

$$X_j = \{x_{j,k}, k \in \mathcal{K}_j\},$$

where $\mathcal{K}_j$ is the appropriate set of location indices for grid points at scale j. We refer to the *vector* of grid locations at level j as $\boldsymbol{x}_j$.

2.3.2 The Unbalanced Haar Transform

The classical Haar transform can be readily extended to non-equidistant grids, even without using the lifting scheme. On a regular grid, the Haar transform considers the input as scaling coefficients $s_{J,k}$ corresponding to scaling basis functions. These basis functions $\varphi_{J,k}$ are indicator functions on the half open interval $[x_{J,k}, x_{J,k+1})$ (as can easily be verified by running the subdivision scheme). For ease of discussion, we work with non-normalized basis functions in this section. Since these intervals all have the same length, the scaling functions are translations of a single *father function*. The resulting function

$$f(x) = \sum_k s_{J,k} \varphi_{J,k}(x)$$

is a piecewise constant, of which Figure 2.12(a) shows a typical detail. The scaling coefficients at the next, coarser scale are then the pairwise averages of $s_{J,k}$. As the figure illustrates, this can be interpreted graphically as averaging two adjacent constant lines into a constant line of double length. The detail coefficient stands for a blocky wavelet correction on this average approximation, where the height of the blocky wavelet is given by the coefficient and its shape (i.e., its interpretation) is of course that of the Haar wavelet function itself.

All this can be immediately generalized to non-equidistant grids, as in Figure 2.12(b). In order to take the irregularity of the grid into account, we now take *weighted averages*. We call

$$I_{j,k} = \int_{-\infty}^{\infty} \varphi_{j,k}(x)dx = x_{j,k+1} - x_{j,k},$$

42   2. Second-generation Wavelets

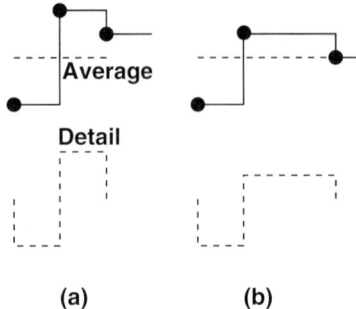

Figure 2.12. The classical and unbalanced Haar transform. The classical Haar transform uses unweighted averages, whereas the unbalanced version weighs the averages according to the lengths of the intervals corresponding to the input values. The resulting detail basis functions have zero integrals on non-equispaced intervals.

the integrals of the basis functions, or the length of the (now non-homogeneous) intervals $[x_{j,k}, x_{j,k+1}]$. The scaling coefficients at the next, coarser scale are then given by

$$s_{j-1,k} = \frac{I_{j,2k} s_{j,2k} + I_{j,2k+1} s_{j,2k+1}}{I_{j-1,k}}.$$

The integrals at coarse scale $I_{j-1,k}$ can be computed from the grid locations, but they also satisfy a trivial two-scale recursion relation

$$I_{j-1,k} = I_{j,2k} + I_{j,2k+1}.$$

The calculation of the detail coefficients does *not* involve weighting. Like in the equidistant case, the detail coefficients are simply

$$w_{j-1,k} = s_{j,2k+1} - s_{j,2k}.$$

Indeed, if the sequence $s_{j,k}$ is constant, we want zero detail values. Weighted details on irregular grids would not be zero for such a constant input, and that is obviously not what we want. A detail coefficient, together with a weighted average, is still sufficient to reconstruct the two original valuess. The detail coefficient can now be interpreted as the height of the "unbalanced" blocky wavelet in Figure 2.12(b). This blocky wavelet again has an integral equal to zero, thanks to the choice of a *weighted average*.

The only difference between the classical Haar algorithm and the unbalanced Haar algorithm is the weighting of the averages. Also, the classical algorithm easily applies to non-equidistant grids. Because of the limited support of a Haar basis function, applying the classical algorithm as if the data were on a regular grid hardly results in an output reflecting the irregular structure of the grid as in the example of Section 2.1. Both classical and unbalanced Haar have the benefits from being orthogonal transforms, but the classical transform on an irregular grid has the disadvantage that it operates with wavelets with nonzero integrals. As further explained in Section 4.4.2, this is an unfavourable situation from a numerical point of view: approximation errors may not converge. This remark has to be nuanced by the fact that we can bring in a weight function

$$w(x) = \sum_{k \in \mathcal{K}_J} \delta(x - x_{J,k}),$$

such that

$$\int_{-\infty}^{\infty} \psi_{j,k}(x) w(x) dx = 0,$$

for the wavelet basis functions $\psi_{j,k}$ of the *classical* (i.e., as opposed to the unbalanced) Haar transform. We can measure all integrals, inner products, and norms with this weight function. This is equivalent to replacing the classical Lebesgue measure with a different measure, in this case the empirical measure corresponding to the observations at hand [29]. The importance of numerical issues, such as convergence of errors, depends on how these errors are measured; see Section 4.4.2 for further discussion on this issue.

Although the main ideas remain the same as in the regular case, the scaling basis functions on irregular point sets are no longer translations of a single father functions: they can be stretched within a single scale. Also, across scales, the scaling basis functions are no longer dyadic dilations. The wavelet functions are still blocky waves, with vanishing integrals, but in order to achieve this zero integral, the Haar function has to be "unbalanced" in height. As a consequence, the classical theory of multiresolution, assuming translations and dyadic dilations, no longer applies.

As illustrated in Figure 2.13, the unbalanced Haar wavelet transform can also be implemented in a lifting version:

$$\begin{aligned} w_{j-1,k} &= s_{j,2k+1} - s_{j,2k} \\ I_{j-1,k} &= I_{j,2k} + I_{j,2k+1} \\ s_{j-1,k} &= s_{j,2k} + \frac{I_{j,2k+1}}{I_{j-1,k}} w_{j-1,k}. \end{aligned}$$

Note that we added the recursive computation of the integrals to the scheme. This is not strictly necessary for this example, as the integral values can be easily retrieved from the grid locations.

This lifting scheme is the generalization of the scheme in Figure 2.2. It is, however, not unique. In some applications, it is useful first to compute the averages, *before* the detail coefficients. This is convenient, for instance, if one seeks to apply a more sophisticated prediction, while keeping simple averaging as the basis for two-scale coarsening. The resulting Haar lifting scheme is that of Figure 2.14. It corresponds to

$$\begin{aligned} s_{j-1,k} &= \frac{I_{j,2k}}{I_{j,2k} + I_{j,2k+1}} s_{j,2k} + \frac{I_{j,2k+1}}{I_{j,2k} + I_{j,2k+1}} s_{j,2k+1} \\ w_{j-1,k} &= \frac{I_{j,2k} + I_{j,2k+1}}{I_{j,2k}} (s_{j,2k+1} - s_{j-1,k}) \\ I_{j-1,k} &= I_{j,2k} + I_{j,2k+1}. \end{aligned}$$

44 2. Second-generation Wavelets

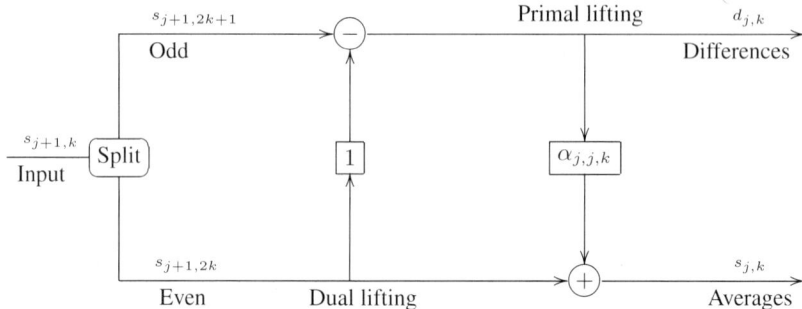

Figure 2.13. First lifting implementation of the unbalanced Haar transform. This is the irregular grid version of Figure 2.2. The update coefficient is $\alpha_{j,k,k} = I_{j+1,2k+1}/I_{j,k}$.

Note that this system requires the introduction of two multiplications by a scalar. Such a rescaling is always immediately invertible, even if it occurs on a single branch (even or odd) of the scheme (as is the case here). This is in contrast to convolutions in update or prediction steps. The Haar transform is exceptional: primal and dual lifting steps contain only simple multiplications, but general (primal and dual) lifting steps have convolutions (filters). Invertibility is then due to the fact that those convolutions take place while keeping the input in a separate branch, as explained in Section 2.2.2.

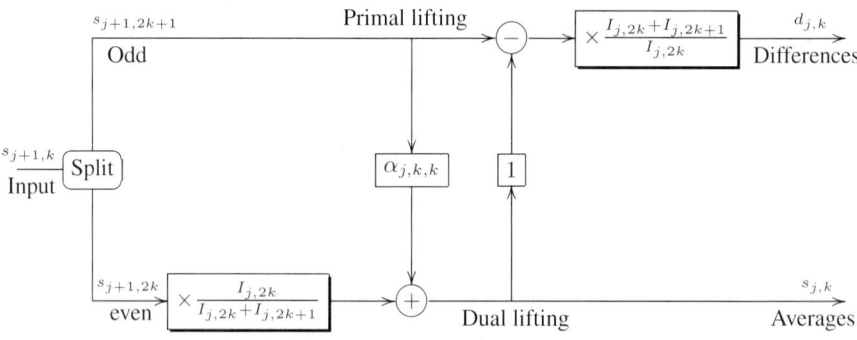

Figure 2.14. Second lifting version of unbalanced Haar transform, this time with update first. This version requires additional rescalings, i.e., multiplications with a scalar.

2.3.3 Prediction Methods

Interpolating Prediction. The introduction of the unbalanced Haar transform in Section 2.3.2 did not rely on lifting. The lifting paradigm, however, turns out to be an excellent tool for the extension of the unbalanced Haar transform towards the construction of general wavelet analyses on irregular point sets.

Indeed, the idea of polynomial interpolation as predictors, discussed in Section 2.2.2, is by no means limited to data on infinite, regular point sets. On an irregular grid, this scheme can be generalized straightforwardly. For the moment, we assume that the interpolation uses the same number of evens to the left and to the right of an odd location. Cubic interpolation, for instance, involves two neighbours on each side. Under this assumption that the number of neighbours is equal on both sides, the degree of the interpolating polynomial is always odd. If the input itself is an exact polynomial of degree lower than this interpolating polynomial, then all detail coefficients are exactly zero. The scaling basis functions have a polynomial reproducing property, independent from the scale: at each scale, for degree p smaller than the degree of the interpolating polynomial, there exists constants $s_{j,k}$ such that

$$x^p = \sum_{k \in \mathcal{K}_j} s_{j,k} \varphi_{j,k}(x). \tag{2.14}$$

The degree of the interpolating polynomial thus determines the number of vanishing moments of the analysis, i.e., the dual number of vanishing moments. This explains why a prediction step is known as *dual lifting*.

As explained in Section 2.2.2, the scaling basis functions resulting from subdivision with interpolating, polynomial prediction are no polynomials or piecewise polynomials, except for the linear case. On the other hand, the basis functions do have the *interpolating property*, i.e., on all scales j, we have

$$\varphi_{j,k}(x_{j,l}) = \delta_{k-l}. \tag{2.15}$$

This interpolating property allows one to evaluate (2.14) in the locations $x_{j,k}$, leading to $s_{j,k} = x_{j,k}^p$; hence

$$x^p = \sum_{k \in \mathcal{K}_j} x_{j,k}^p \varphi_{j,k}(x). \tag{2.16}$$

This holds not only for polynomials, but for all functions $f_j(x)$ spanned by the scaling basis

$$f_j(x) = \sum_{k \in \mathcal{K}_j} f_j(x_{j,k}) \varphi_{j,k}(x).$$

Recall that the first interpretation of coefficients in a (scaling function) basis is how much that particalur (scaling) basis *function* contributes to the synthesis of the target function $f_j(x)$. In this particular case, those coefficients have a second interpretation, namely the *value* of that function $f_j(x)$ in a particular point, $x_{j,k}$. The prediction is based on this second interpretation. If we have observations f_k and use them as scaling coefficients at the finest scale $s_{J,k} = f_k$, we are implicitly assuming that the observations come from a function in the space spanned by these scaling functions. This is probably an approximation, since it is unlikely that in a given application we are dealing with a function in that function space. But with interpolating scaling functions, using observations as scaling coefficients is less dramatic than in general. Indeed, in general, this approximation not only assumes that we are working with a function inside the space spanned by the scaling functions, it also assumes that we observe scaling coefficients, rather than function values, which is quite unrealistic. Strang and Nguyen call this the *wavelet crime* [73, page 232].

Near boundaries of a finite interval, this interpolating prediction has to be modified simply because, on one side, the boundary limits the number of available neighbours. There are several options to deal with this. The first one keeps the same number of evens in the construction of a prediction polynomial. This implies that the locations of these evens are asymmetric around the location of the odd: less evens on the side of the boundary and more on the other side. The strategy of choosing an equal number on both sides is locally replaced with a strategy of choosing the nearest evens. We could also apply this nearest neighbours strategy for all predictions. In most cases, one can expect the nearest even neighbours to be distributed equally on both sides of an odd point, but this is certainly no longer the case for highly irregular grids with wide gaps next to densely observed intervals. A nearest neighbours strategy as a boundary treatment has some important numerical problems, as explained in Section 4.4.4. A second option is keeping the same number on the interior side of the odd point (i.e., half of the maximum possible number of evens when there is no boundary limitation) while staying with a limited number of even neighbours on the boundary side. This reduces the number of vanishing moments near the boundary. That number is reduced even further in a third option, where the distribution of even neighbours is symmetric over the two sides by limiting the number on the interior in accordance with the available neighbours on the boundary side.

Average Interpolating Prediction. Polynomial interpolating prediction with a symmetric choice of neighbours around an odd point always leads to an even number of dual vanishing moments. A simple example of a lifting scheme with an odd number of dual vanishing moments can be obtained by first pairwise averaging the evens and odds, followed by a prediction based on those averages, rather than on the even inputs. This allows us to consider an odd number of those averages, with in the middle the average of the odd in the location where we want to use the prediction and its even neighbour to the left.

We impose that the scaling coefficients $s_{j,k}$ can be interpreted, not only as contributions in the basis decomposition

$$f_j(x) = \sum_{k \in \mathcal{K}_j} s_{j,k} \varphi_{j,k},$$

but also as the *average function value* of $f_j(x)$ on the half-open interval $[x_{j,k}, x_{j,k+1})$, i.e.,

$$s_{j,k} = \frac{\int_{x_{j,k}}^{x_{j,k+1}} f_j(x) dx}{x_{j,k+1} - x_{j,k}}.$$

2.3 The Construction of Second-generation Wavelets

Since the basis functions $\varphi_{j,k}$ themselves are represented by a Kronecker sequence of scaling coefficients, their averages are one on the corresponding interval $[x_{j,k}, x_{j,k+1})$ and zero on all other level-j intervals. The function f_j is said to be *average interpolating*.

The Haar transform satisfies this assumption. Indeed, if $f_j(x)$ is spanned by the scaling functions at level j, it is a left-continuous piecewise constant function with knots $x_{j,k}$. The coefficients are also the averages on each interval, since the basis functions are indicator functions. The Haar decomposition *also* satisfies the interpretation used in interpolating prediction: the coefficients are also function values in the knots. Indeed, the function values in the knots must be equal to the averages on the half-open intervals containing them, since f_j is constant on all these intervals. Satisfying both assumptions together is, of course, exceptional.

As before, we call

$$I_{j,k} = x_{j,k+1} - x_{j,k}$$

the length of the interval $[x_{j,k}, x_{j,k+1})$. Because the basis function $\varphi_{j,k}$ is average interpolating with coefficients the Kronecker sequence, it has average value equal to one on this interval and zero averages on all level j intervals outside $[x_{j,k}, x_{j,k+1})$. For the integral, this means that

$$\int_{-\infty}^{\infty} \varphi_{j,k}(x) dx = I_{j,k}.$$

We now show that imposing this average interpolation property on the coefficients implies that coarse-scale scaling coefficients are indeed weighted averages of two, adjacent fine-scale scaling coefficients. Recall that we proposed this averaging as a way to have an odd number of dual vanishing moments with a symmetric prediction scheme. By $\mathcal{V}_j$ we denote the function space spanned by the scaling basis functions at level j. All functions $f_j \in \mathcal{V}_j$ share the property that their coefficients in the scaling decomposition have the meaning of averages on the level j intervals. A function $f_j \in \mathcal{V}_j$ not only has a decomposition in $\mathcal{V}_j$, but also in $\mathcal{V}_{j+1} \supset \mathcal{V}_j$:

$$f_j(x) = \sum_{k \in \mathcal{K}_j} s_{j,k} \varphi_{j,k}(x) = \sum_{k \in \mathcal{K}_{j+1}} s_{j+1,k} \varphi_{j+1,k}(x),$$

for which we can state that

$$s_{j,k} = \frac{1}{I_{j,k}} \int_{x_{j,k}}^{x_{j,k+1}} f_j(x) dx$$

$$s_{j+1,2k} = \frac{1}{I_{j+1,2k}} \int_{x_{j+1,2k}}^{x_{j+1,2k+1}} f_j(x) dx$$

$$s_{j+1,2k+1} = \frac{1}{I_{j+1,2k+1}} \int_{x_{j+1,2k+1}}^{x_{j+1,2k+2}} f_j(x) dx.$$

Since we have that $x_{j,k} = x_{j+1,2k}$ and $x_{j,k+1} = x_{j+1,2k+2}$, it follows that, for any function in $\mathcal{V}_j$

48   2. Second-generation Wavelets

$$s_{j,k} = \frac{I_{j+1,2k} \cdot s_{j+1,2k} + I_{j+1,2k+1} \cdot s_{j+1,2k+1}}{I_{j,k}}. \quad (2.17)$$

In other words, the scaling coefficients at scale j are indeed found as pairwise averages of the coefficients at scale $j+1$.

The scheme for this decomposition appears in Figure 2.15. It is an extension of the update-first lifting implementation of the unbalanced Haar transform in Figure 2.14. Recall that interpolating subdivision was an extension of the prediction-first implementation of the same unbalanced Haar transform. The first update step,

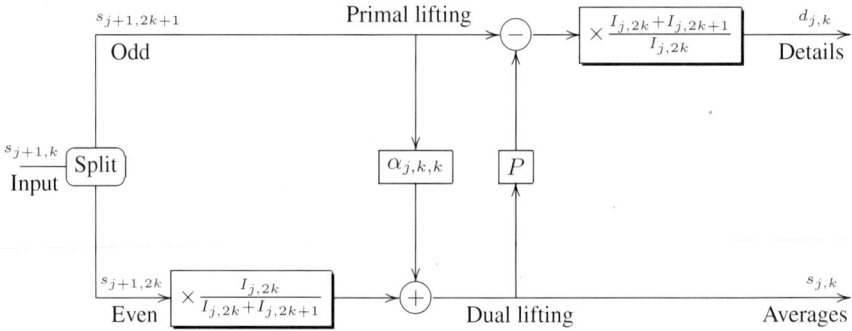

Figure 2.15. Average interpolating prediction. In order to have the correct interpretation for the scaling coefficients, namely that they are averages of the underlying function on the intervals $[x_{j,k}, x_{j,k+1})$, we first have to include a Haar update step. The shaded boxes in the scheme stand for a multiplication with a scalar. This is in contrast to the rectangular boxes, which stand for convolutions (filtering). In the Haar transform of Figure 2.2, these convolutions exceptionally reduce to a simple rescaling, but this is not generally true. The main difference between a rescaling and a convolution lies in the fact that the former is readily invertible, even if it takes place on the even or odd branch itself. A convolution in a lifting scheme is only invertible if it takes place on a branch *connecting* the main branches (even and odd), so that the input of the convolution operation is still available afterwards.

i.e., the average computation, was motivated by the interpretation we want to give to the coefficients, namely that the coefficients $s_{j,k}$ are equal to the averages of the underlying function $f_j = \sum_k s_{j,k} \varphi_{j,k}$ on the intervals $[x_{j,k}, x_{j,k+1})$. A side consequence of this is that the wavelet functions have a vanishing integral, whatever the exact predictor may be. Indeed, because of (2.17), we have that the averages over all scaling coefficients at one level are a constant over all levels:

$$\sum_{k \in \mathcal{K}_j} I_{j,k} \cdot s_{j,k} = \sum_{k \in \mathcal{K}_{j+1}} I_{j+1,k} \cdot s_{j+1,k}.$$

Now, if we run subdivision with $s_{j,l} = 0$, for all l and $w_{j,l} = \delta_{k,l}$, the average of scaling coefficients at all subsequent levels $J \geq j$ will be zero, and so will be the integral of the limiting basis function. If the prediction step is followed by a second

2.3 The Construction of Second-generation Wavelets

update step, that update step changes the wavelet basis functions, and so it may undo this first (free) vanishing moment, if it is uncarefully designed.

Note that after applying a *forward* step on a *fine*-scale function $f_{j+1} \in \mathcal{V}_{j+1}$, the *coarse*-scale coefficients $s_{j,k}$ are *not* (necessarily) the averages of that fine-scale function on the coarse-scale intervals $[x_{j,k}, x_{j,k+1})$. Indeed, we have $f_{j+1} = f_j + g_j$, with

$$f_j = \sum_{k \in \mathcal{K}_j} s_{j,k} \varphi_{j,k}$$

$$g_j = \sum_{k \in \mathcal{K}'_j} w_{j,k} \psi_{j,k},$$

and the averages

$$\int_{x_{j,k}}^{x_{j,k+1}} \psi_{j,k}(x) dx$$

of the wavelets $\psi_{j,k}$ on the coarse-scale intervals $[x_{j,k}, x_{j,k+1})$ are not necessarily zero, even if these wavelet functions have a zero integral

$$\int_{-\infty}^{\infty} \psi_{j,k}(x) dx = 0$$

on the entire real axis.

The details of the predictor can be filled in by requiring that if the input consists of observations from a polynomial up to degree n say, then all detail coefficients are zero. This means that all (piecewise) polynomial data, or data that can be well approximated by (piecewise) polynomials, achieve high compression rates through this decomposition. If the decomposition of x^p has all details equal to zero, this implies that $x^p \in \mathcal{V}_j, \forall j$. The polynomial must be in the space spanned by the scaling functions at (any) scale j. As a consequence, the scaling coefficients of that polynomial are the averages of the polynomial on the level j intervals. Suppose now that we are given $n+1$ adjacent averages, where $n = 2m$ is an even number. These averages correspond to the scaling coefficients $s_{j,k-m}, \ldots, s_{j,k+m-1}$ with the knot locations $x_{j,k-m}, \ldots, x_{j,k+m}$. There is always a polynomial $p_{j,k,n}(x)$ of degree n that fits those averages. If we call

$$P_{j,k,n}(x) = \int_{x_{j,0}}^{x} p_{j,k,n}(u) du,$$

we find $n+1$ interpolation conditions for $P_{j,k,n}(x)$:

$$P_n(x_{j,k-m+1}) = s_{j,k-m} I_{j,k-m}$$

$$\ldots$$

$$P_n(x_{j,k+m}) = \sum_{l=-m}^{m-1} s_{j,l} I_{j,l}.$$

We now define the detail coefficients at scale $j+1$ as

$$d_{j,k} = s_{j+1,2k+1} - \frac{P_{j,k,n}(x_{j+1,2k+1}) - P_{j,k,n}(x_{j+1,2k})}{I_{j+1,2k+1}}. \qquad (2.18)$$

Expressions (2.17) and (2.18) constitute an average interpolating lifting scheme. The first, primal, step, i.e., expression (2.17), ensures that the scaling coefficients at all scales can be interpreted not only as contributions in a scaling basis decomposition, but also as averages of the scaling approximation at that scale on the intervals at that scale. The second, dual, step, i.e., expression (2.18), ensures that all polynomials of a certain degree can be represented without detail coefficients.

The structure of the grid on which the data live shows up on three occasions:

1. upon weighting the coarse-scale average calculations in the primary primal lifting step;
2. upon establishing the average interpolating, predicting polynomial;
3. upon evaluating that polynomial on the fine-scale interval before subtracting that evaluation from the observed value.

Unlike the polynomial predicting scheme, the first two occasions do not rely on the *locations* of the observations, but rather on the *intervals*. The input is interpreted as *averages* over fine-scale intervals. If we apply this scheme and take just observations as input, we are fully guilty of the wavelet crime, as explained in the section on interpolating prediction. The excuses we could use in that interpolating prediction case do not apply here. As in Section 2.3.2, we can redefine the length of an interval by measuring it using an empirical measure [29]:

$$I_{j,k} = \int_{x_{j,k}}^{x_{j,k+1}} \sum_{l \in \mathbb{Z}} \delta(x - x_{J,l}) dx = \#\{x_{J,l}, l \in \mathbb{Z} | x_{j,k} \leq x_{J,l} < x_{j,k+1}\}.$$

At the finest scale, the averages over the intervals now coincide with the observations, and we can safely use these values as input. The price we pay is loss of adaptivity in two out of the three occasions where the grid plays a role, namely whenever the grid structure shows up through the lengths of the intervals: the unbalanced Haar averaging is replaced with unweighted, plain Haar averaging and the average predicting polynomial is now based on straightforward, unweighted averages of the observations involved. The impact of this empirical weighting on issues like smoothness and stability is an interesting subject of research.

The average interpolating scheme in this text was based on computing the weighted averages as in an unbalanced Haar transform. It is also possible to start with a full Haar decomposition step and modify the subsequent prediction step accordingly. By doing so, we get a scheme with three lifting steps: initial Haar update, initial Haar prediction and modified average interpolating prediction. It is then possibe to switch update and prediction within the initial Haar decomposition. The resulting scheme consists of three alternating lifting steps: Haar prediction, Haar update and average interpolating prediction. It is a nice exercise to design this modified prediction operation; the solution can be found in the literature [75].

Other Predictors. The idea of interpolation prediction can easily be extended beyond polynomials. Section 4.4.4 explores so-called natural interpolation, a particular case of piecewise rational interpolation in two dimensions. Other predictors replace interpolation with least squares, also discussed in Section 4.4.4. The limiting functions from subdivision with this predictor are highly non-smooth, especially if the least squares prediction does not weigh the importance of the even points according to their distance to the odd point where the prediction is being evaluated.

For more general dual lifting steps, there is not always an immediate interpretation of the even and odd input values, as was the case for interpolating and average interpolating prediction. If the input values cannot be interpreted as function values at a given location, or as averages over an interval, then one stays with the interpretation as coefficients in a fine-scale scaling basis decomposition. The prerequisites for some applications may be best expressed directly in terms of those basis functions. In computer graphics, for instance, cubic splines, i.e., piecewise cubic polynomials that are continuously differentiable in the knots, are particularly interesting, because of their nice C^2 continuity properties, in combination with an easy manipulation through the knots. Cubic B-splines constitute a basis for this class of functions. These basis functions $\varphi_{j,k}(x)$ satisfy a two-scale equation

$$\varphi_{j,k} = \sum_{l \in \mathbb{Z}} h_l \varphi_{j+1,l-2k},$$

with $h_k = 1/8(1,4,6,4,1)$. As explained in Section 2.2.4, the solution of a two-scale equation can be found (at least numerically) by running a subdivision scheme on the input

$$\varphi_{j,k} = \sum_{l \in \mathbb{Z}} \delta_{l-k} \varphi_{j,l}.$$

In other words, the input of the subdivision is a Kronecker sequence of coefficients associated with scaling functions at level j. After one step, we have the two-scale equation coefficients, corresponding to scaling functions at level $j + 1$. In the next steps of the refinement process we express $\varphi_{j,k}$ as a combination of ever finer basis functions, where the coefficients gradually converge to the values of $\varphi_{j,k}$ itself. This subdivision scheme can be decomposed into lifting steps, leading to the following forward lifted wavelet transform:

$$\begin{aligned} s_{j,k} &= 2s_{j+1,2k} - \frac{1}{2}(s_{j+1,2k-1} + s_{j+1,2k+1}) \\ d_{j,k} &= s_{j+1,2k+1} - \frac{1}{2}(s_{j,k} + s_{j,k+1}). \end{aligned}$$

This lifting scheme consists of two steps, beginning with an update step. Unlike in the case of average interpolation, the update cannot be interpreted in terms of averaging, nor does it make the detail basis functions have zero integrals. As a consequence, we need at least one more update step to generate wavelets with zero integral. The consequent prediction step looks a bit like linear polynomial prediction, except that the even neighbours $s_{j+1,2k}$ and $s_{j+1,2k+2}$ have already been updated and are replaced by $s_{j,k}$ and $s_{j,k+1}$. Anyway, the example illustrates that far from all lifting transforms can be readily interpreted as a series of prediction and update steps. Also note that we are constructing B-spline *basis* functions, and this is *all but* equivalent to using (interpolating) splines as prediction. Recall that interpolating polynomial prediction (discussed at the beginning of this section) did not lead to polynomial basis functions either.

Since we may not be able to *interpret* all lifting steps that lead to a desirable multiresolution analysis and wavelet decomposition, the question arises as to how we can find the lifting implementation of the cubic spline subdivision scheme. As mentioned in Section 2.2.6, there exists a procedure to decompose every existing wavelet transform into lifting steps.

2.3.4 Updates for Vanishing Moments

As explained in Section 2.2.3, running one step of the subdivision scheme reveals the coefficients in the two-scale equation (2.8). Taking the integrals or higher order moments on both sides leads to

$$M_{j,k}^{[p]} = \sum_{l} h_{j,k,l} M_{j+1,l}^{[p]}, \qquad (2.19)$$

where

$$M_{j,k}^{[p]} = \int_{-\infty}^{\infty} x^p \varphi_{j,k}(x) dx.$$

The calculation of these moments is necessary when the update is designed to generate wavelets with vanishing moments. In the classical, regular setting, computation of these moments can be based on the fact that all scaling functions are dilations and translations of a single basis function. This is no longer the case if the data live on an irregular grid. The scaling function moments at scale j can be computed from the moments at (finer) scale $j + 1$. Expression (2.19) can be seen as one step of a *forward* wavelet decomposition: it computes coarse-scale moments from fine-scale moments. The coefficients of this forward transform are, however, those that appear in the two-scale equation for the scaling *basis functions*, normally used in the synthesis, i.e., the inverse transform or reconstruction step, not for the calculation of the coefficients (analysis). The moments are thus computed by running a forward wavelet transform with inverse wavelet coefficients. Unless the transform (matrix) is orthogonal, these inverse coefficients are different from the forward coefficients. The inverse coefficients are, however, easy to find in each step by running one step of the subdivision scheme at that scale, as explained in Section 2.2.4.

A slightly more serious problem originates from the *forward* nature of these moment calculations. We also need the moments upon reconstruction, when we are proceeding from coarse to fine scales. Since the moments are calculated from fine to coarse, the inverse second-generation wavelet transform starts with a forward transform to find the moments.

As soon as we have the moments of the scaling functions, we can impose that the primal wavelets have zero moments. Indeed, from Expression (2.9), we know that

$$\psi_{j,k} = \varphi_{j+1,2k+1} - \sum_l \alpha_{j,k,l} \varphi_{j,k},$$

from which we can impose that:

$$0 = M^{[p]}_{j+1,2k+1} - \sum_l \alpha_{j,k,l} M_{j,k}.$$

This is a set of equations for the update coefficients $\alpha_{j,k,l}$.

This update preserves the average value (DC-component) in the set of scaling coefficients. Indeed, integrating the equality

$$\sum_{k \in \mathbb{Z}} s_{j+1,k} \varphi_{j+1,k} = \sum_{k \in \mathbb{Z}} s_{j,k} \varphi_{j,k} + w_{j,k} \psi_{j,k}$$

leads to

$$\sum_{k \in \mathbb{Z}} s_{j+1,k} M_{j+1,k} = \sum_{k \in \mathbb{Z}} s_{j,k} M_{j,k},$$

using the fact that the wavelet basis functions have vanishing integrals. On a regular grid, using the correct normalization, the equation above reduces to stating that the DC-component is preserved throughout scales. This condition is sometimes taken as initial guidline in the design of update steps [75].

2.4 Lifting in Two (and More) Dimensions

Although the extension of primal and dual lifting steps into higher dimensions may require some considerable effort, the first obstacle in the design of a 2-d lifting scheme is probably in the splitting stage. Indeed, in a one-dimensional (1-d) lifting scheme, the splitting step determines which input values are going to be predicted and which neighbours are going to be used for that prediction. The notion of "neighbours" is defined by "left" and "right" — i.e., "previous" and "next" — input values. In two dimensions, as in one dimension, we could rather define neighbours of one observation as the "k nearest" other observations, for some constant k, i.e., the k observations with the smallest (Euclidian) distance to the point in consideration. The problem with this (geometric) definition is that in one dimension all "nearest" points might fall on one side of the point of interest (namely, if there is large gap without observations on the other side). In two dimensions, there are not just two sides for each observation, but an infinite number of possible search directions. The question is how to define neighbourhood in each possible direction, and how to distingiush a sparsely observed area from the boundary of the domain of observations. An alternative could be the so-called *ball neighbourhood*, where a neighbour is defined as an observation that lies within a given distance of the point of interest. This immediately poses the question of how large that distance should be chosen. The choice is probably going to be space and scale adaptive. If the data are on a regular, tensor-product grid, it seems straightforward to define first-order neighbours as the four points north, east, south and west of the point of interest. Second-order neighbours may then include the four points in diagonal directions. For irregular grids, neighbourhood can be defined through a triangulation of the observations.

2.4.1 Definitions and Construction of Triangulations

Given a set of b planer data points v_i, with $i = 1, \ldots, b$, a polyline is a curve consisting of the line segments $[v_i v_{i+1}] := \{\alpha v_i + (1 - \alpha) v_{i+1} | 0 \leq \alpha \leq 1\}$, for $i = 1, \ldots, b - 1$. The polyline is called *closed* if it also contains the segment $[v_b v_1]$. Consider now a set $V = \{v_i\}_{i=1,\ldots,N} \subset \mathbb{R}^2$ of fixed planar data points; for instance, in many applications, those points are the locations of observations. These points are refered to as *vertices* (singular: vertex). A boundary $\partial V \in V^b$, for some integer $3 \leq b \leq N$ is an ordered b-tuple, such that the closed polyline has no self-intersections and such that the simply connected polygon $\Omega_{\partial V} \subset \mathbb{R}^2$ circumscribed by this polyline contains all points in V: $V \subset \Omega_{\partial V}$. This boundary is not unique. It could be the convex hull, but in many applications it makes more sense to work on a smaller domain of data.

A *triangulation* or a *triangular mesh* of V, $\mathcal{T}(V) = \{T \in V^3\}$ is a complete partitioning of such a circumscribing polygon Ω into triangular regions. Each triangle is defined as a triple of vertices from V. The corresponding triangular region is the open set of points within the triangle's sides. All triangular regions must be mutually exclusive and the union of their closures must equal Ω (which we assume to be closed). Each pair of vertices within a triangle constitutes an *edge*. Two vertices are called (immediate or first-order) *neighbours* if there is an edge (or, equivalently, a triangle) that contains both.

2.4.2 Delaunay Triangulation

Obviously, for a given set of observations, there are many triangulations possible and not all of them have the same quality in expressing the neighbourhood structure of the observations (which was our first objective). We therefore look for a triangulation that describes that structure in some optimal way.

A *Voronoi tesselation* of V is a maximal subdivision of Ω into Voronoi cells, i.e., subsets Ω_i according to

$$\Omega_i = \{x \in \Omega | \forall v_j \in V, j \neq i : d(x, v_i) \leq d(x, v_j)\},$$

where $d(\cdot, \cdot)$ is a distance function. The remainder of this section adopts the Euclidian distance. In words, a Voronoi tesselation assigns every planar point to the vertex that is nearest to that point. A Voronoi cell is the polygonal region containing all the points assigned to a given vertex (including that vertex itself). Whether points on the intersection of two neighbouring cells are assigned to one or the other cell is not important for this discussion.

Associated with a Voronoi tesselation is a Delaunay triangulation: two vertices v_i and v_j are connected by an edge in a Delaunay triangulation if their Voronoi cells have a common boundary of nonzero length. Pathological cases, where more than three cells meet in a single point, can be treated separately in an arbitrary way. Among many others, a Delaunay triangulation has the following properties:

Property 1 (empty circumcircles) *Every circumcirle of a triangle has no other vertex inside.*

Property 2 (maximin angle) *A Delaunay triangulation is a triangulation with the largest possible minimal angle.*

If $\Omega = \mathbb{R}^2$, then the boundary of a Delaunay triangulation coincides with the convex hull of the set of vertices (not taking into account edges connecting a vertex with infinity).

The construction of a Delaunay algorithm on $\mathbb{R}^2$ proceeds vertex by vertex: every time a vertex is inserted and connected, a local search is performed to find and solve violations of the Delaunay properties. A triangulation of a non-convex subset $\Omega \subset \mathbb{R}^2$ can be obtained from a full triangulation of $\mathbb{R}^2$ by removing all triangles outside Ω. In some applications, an automatic construction of a non-convex Delaunay triangulation might be interesting. A criterion could be that vertices are only connected if their Voronoi cells meet inside the convex hull of the set of vertices. This corresponds to removing all triangles on the boundary with obtuse inside angles. If this procedure reveals a new triangle with obtuse inside angle, the criterion should be applied in an interative way. This triangle removal should, however, make an exception for vertices that have two or three obtuse angles in two or three triangles on the boundary, since otherwise the triangulation becomes disconnected in those vertices; whenever two or three obtuse angles share a common vertex, the policy followed here is to remove the triangle with the largest angle in that point only. Instead of removing boundary triangles with obtuse inside angle, the criterion could also be relaxed; for instance, the threshold could be at inside angles larger than $2\pi/3$ or so.

Figure 2.16 has an example of the effect of this triangulation, compared to the result of a plain Delaunay approach.

2.4.3 Multiscale Triangulations

Since we are using triangulation as a tool to define a neighbourhood, and since we use the neighbourhood structure at every resolution level with the vertices at that level, we have to introduce a multilevel triangulation. In order to define a multiscale triangulation, we first construct a sequence of nested subsets of vertices, i.e., by $V_j \subset V$ we denote the set of vertices at level j and let $\mathcal{T}(V_j)$ be a triangulation of this subset. The construction of this sequence of nested subsets is, of course, not

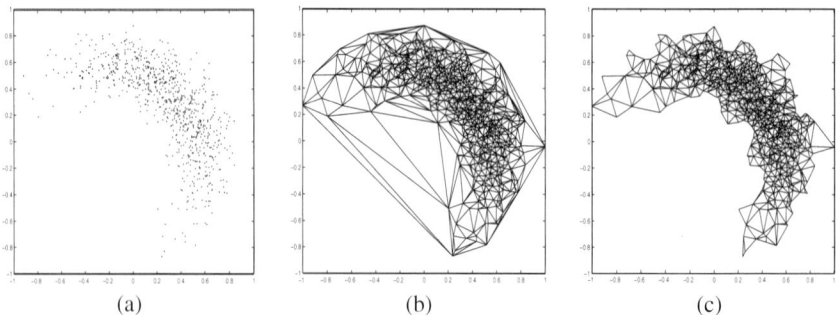

Figure 2.16. In some applications, samples are taken from a non-convex area (a). If the exact contour is unknown, it is interesting to have a triangulation which automatically seeks for an appropriate boundary. The boundary of a Delaunay triangulation always coincides with the convex hull (b). Removing narrow triangles near the boundary yields the result in (c).

unique. In the example explored below, descending one level corresponds to taking out a single vertex.

The construction of a multilevel mesh can be seen as an example of a procedure known in computer graphics literature as *mesh simplification*. It may form part of a *remeshing* strategy, i.e., a procedure to reconstruct a different mesh for a given set of observations that is easier to deal with. A typical example is a remeshing towards a mesh with *subdivision connectivity*. A mesh with subdivision connectivity is a mesh which can be obtained by refining a coarse mesh in a quasi-regular way: in each refinement step, every edge is subdivided into two edges by insertion of a new vertex. Since every triangle consists of three edges, there are three new vertices associated with every triangle. Those three new vertices within each triangle are mutually connected. Since every new vertex is associated with an edge between two coarse-scale triangles, this procedures makes every new vertex connect to exactly six other vertices: two coarse-scale vertices ("evens") and four fine-scale vertices ("odds"). The refinement within each triangle of an irregular coarse-scale triangulation can also be completely regular; every existing edge is then subdivided into two edges of *equal length*. The resulting mesh is then called *semi-regular*; although the coarse-scale mesh may have an arbitrary structure, the vertices added in finer levels have a fixed connectivity and all subtriangles of an original triangle in the coarse triangulation have the same shape. Meshes that come from observed data, such as a Delaunay triangulation on scattered data, very probably do not have subdivision connectivity. Yet, subdivsion connectivity is a necessary condition for many multiscale routines, including wavelet transforms (multiscale decompositions). Mesh simplification is then a first step to obtain a coarse-scale triangulation, from which a new, semi-regular fine-scale mesh can be constructed. Obviously, it is far from guaranteed that the original vertices are located such that the new fine-scale mesh uses exactly the same set of vertices, without leaving some of the original vertices disconnected. In other words, many remeshing procedures choose new vertices. This defines a new triangular surface, for which the error (compared with the input) has to be kept under control.

In some applications, such as smoothing, a remeshing with new points at locations different from the original observations destroys the input statistics (such as the typical assumption that the observations are statistically independent) and, therefore, not an interesting tool. Mesh simplification then serves as a tool to generate a multilevel mesh, where the fine-scale triangulation corresponds to the input mesh. As an example, we discuss a multiscale Delaunay triangulation: at every level, all triangles satisfy the Delaunay conditions w.r.t. the vertices at that level. Maintaining the Dealunay properties throughout the subsequent levels is an interesting strategy if the objective is to use the resulting multilevel grid directly in a wavelet decomposition. It is probably far less interesting if the eventual goal is to obtain a coarse-scale grid from which a remeshing routine should generate a semi-regular mesh.

2.4.4 Multiscale Delaunay Triangulations

In this discussion we assume that, when a vertex is removed, the locations of the remaining vertices do not change. This is not at all an absolute requirement: the exact point location associated with wavelet coefficients may shift throughout the levels, for instance, if one assigns the average position of evens and odds to the resulting scaling coefficients. We have not done this in the examples discussed so far, but it might be useful in some interpretations of the coefficients.

The decomposition of the data on a triangular grid into a multiscale representation requires a local retriangulation near the vertices that are processed and taken out when going from a resolution level j to a next, coarser level $j+1$. The algorithm assumes that neighbours are never removed within the same step. This is true in the case of a classical (univariate or bivariate tensor-product) wavelet transform: the subsampling corresponds to splitting into observations with even and odd index. In the case of scattered data, scale is a continuous notion: every data point lives on its own scale, depending on the distance to its neighbours. As illustrated in Chapter 4, mixing these scales into a single level of resolution may lead to unstable transforms. Selecting an appropriate set of non-adjacent vertices for subsampling an irregular grid would be a non-trivial task anyway. Therefore, a typical procedure selects just a single vertex in each resolution step, for instance the vertex living on the smallest Voronoi cell. The procedure then computes a wavelet coefficient from the value in this point and its neighbours before leaving it out of the triangulation.

Vertices away from the Boundary. Consider an inside (non-boundary) vertex $v_0 = (x_0, y_0) \notin \partial V$ and its set of neighbours $\mathcal{N}(v_0) \subset V$. When v_0 is left out, it suffices to retriangulate the polygon inside the edges between these neighbours, as follows from these two observations:

Observation 2.1 *If the circumcircle of a triangle includes no other sample points, this triangle must be part of the Delaunay triangulation. (This is the opposite formulation of property 1.)*

Proof: Consider the centre of the circumcircle. Its distance to the three points is equal to the radius. Since no other sample point lies within this circle, no other point can have a smaller distance, and so, this centre is where the three Voronoi cells of the three sample points meet. Since their Voronoi cells are adjacent, the three points are connected by edges in a single triangle.

Observation 2.2 *Adding a vertex to a Delaunay triangulation leaves all existing triangles intact, except for those consisting of three neighbours of the new vertex in the new triangulation.*

Stated otherwise: if you remove this newly added vertex, the effect remains within the polygon of neighbours. Indeed, suppose v_0 is added outside the circumcircle of a given triangle. If no other vertex lies within this circumcircle, then this triangle must be part of the Delaunay triangulation, whether or not v_0 has been added.

The local retriangulation of the polygon of neighbours can then be done in a constructive way, using the following observation:

Observation 2.3 *Any triangulation of a polygon without involvement of internal vertices must contain at least one triangle of three successive vertices on the boundary.*

Since we want to retriangulate a hole of which the boundary edges for sure remain part of a Delaunay triangulation, we can find a Delaunay triangulation that exactly fills the hole without overlapping the other triangles. As a consequence, we can just look for a triple of adjacent vertices with no other vertex in its circumcircle, add the triangle with these vertices to the current triangulation and proceed with the polygon obtained by removing the middle of these three adjacent vertices.

Vertices on the Boundary. If a vertex v_0 on the boundary is left out, then the objective is to find a local retriangulation which is consistent with the initially adopted rules of triangulation, i.e., to find the global triangulation as if this were the initial level of resolution, just using a local procedure.

The set of neighbours now constitutes a non-circular polyline. The algorithm first identifies the end points, say v_n and v_m, of this chain of neighbours, i.e., $v_n, v_m \in \mathcal{N}(v_0)$. Next, it finds the segment between v_n and v_m of the convex hull of the remaining vertices V_j at level j. We do this because we know that the full Delaunay triangulation fills up convex regions. The third step of the boundary correction is to triangulate between that convex hull segment and the remaining triangulation after taking away v_0 and the triangles containing this vertex. Finally, the convex hull segment is inspected for obtuse triangles and those triangles are removed, just as in the initial triangulation procedure at the finest scale. This retriangulation operates locally within the polyline of neighbours of v_0. As Figure 2.17 illustrates, some vertices, or even whole parts of the triangulation, may become isolated or connected to the rest through a single vertex only. This mostly happens near the corners of the configuration. An additional step of the algorithm locally scans the triangulation for such phenomena and then bridges the gaps with a minumum of additional (obtuse) triangles. If more than one "bridge" is possible, then the algorithm choses the solution with the smallest obtuse triangle.

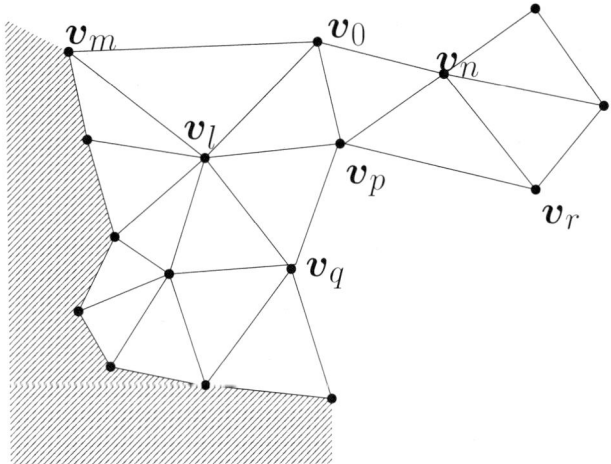

Figure 2.17. Local retriangulation after removing a vertex on the boundary. Suppose that at a given stage in the multiscale triangulation, it is decided to take out vertex v_0. The triangulation depicted is the right upper corner of a larger configuration, indicated by the shaded area. The algorithm first identifies the end points v_n and v_n of the polyline connecting the neighbours $\mathcal{N}(v_0) = \{v_m, v_l, v_p, v_n\}$. Next, it looks for the convex hull of these neighbours, which can always be seen as a pair of polylines connecting v_m and v_n. The procedure picks the one that lies on the outside with respect to the triangulation. In this example, that polyline is simply the line connecting the two end points. It serves as a preliminary boundary segment after the removal of v_0. The region bounded by this convex hull segment and the original polyline of neighbours in $\mathcal{N}(v_0)$ is then triangulated in the third step. Finally, all obtuse boundary triangles are taken away from this local triangulation. In this example, all newly added triangles are taken away. As a consequence, the global triangulation falls apart into two fragments, only connected through a single vertex v_p. Therefore, this vertex has to be bridged by adding the triangle $\Delta v_p v_q v_r$, which has the smallest (still obtuse) angle in v_p of the two possible "bridges" (the other possibility being $\Delta v_p v_s v_n$).

3. Nonlinear and Adaptive Lifting

In the previous chapter we introduced the lifting scheme, both as a way to implement the wavelet transform using in-place computations and as a way to extend the wavelet transform to more general cases, e.g. data defined on irregular grids. In particular, we discussed the unbalanced Haar transform and transforms based on polynomial interpolation schemes. In this chapter we deal with opportunities the lifting scheme yields to build transforms that also take the data (the values of the data) into account instead of only its underlying grid.

3.1 Nonlinear Filters

Before introducing lifting schemes that are data driven in the sense that the choice for primal and/or dual lifting is fully determined by the values of the data, we consider an extension of prediction and update operators.

Until this stage we only considered classical linear filters to be used for prediction and updating, e.g., interpolation filters. A closer look at the lifting scheme shows that all operations in the scheme are not assumed necessarily to be linear, at least if we only consider the scheme itself. Of course, relations with classical MRA and wavelet basis functions, as described in the previous chapter, may not hold anymore in different settings.

A study by Goutsias and Heijmans [41] demonstrated how morphological operators can be used within the lifting scheme to build nonlinear wavelet transforms. As a particular case of nonlinear lifting, we start this chapter by considering the max-lifting scheme see also [41].

3.1.1 The Max-lifting Scheme

The max-lifting scheme can be seen as the result of a natural need to be more flexible when using the Haar transform. We recall that for this transform we have

$$s_{j,k} = \frac{s_{j+1,2k} + s_{j+1,2k+1}}{2}$$
$$d_{j,k} = s_{j+1,2k+1} - s_{j+1,2k}.$$

So the Haar transform generates a detail signal by subtracting even entries of the data from the odd ones. This procedure has been visualized in Figure 3.1 for two block functions. The sampled block functions in both columns are totally the same up to a shift over once. The result after prediction and updating one time has been depicted in Figure 3.1.(b) as (d_j) and in Figure 3.1.(c) as (s_j). Clearly, the detail signal in the left column does not show the jumps of the block functions as does the detail signal in the right column. The reason for this is that odd and even entries are interchanged when comparing the two datasets.

To make the Haar transform more robust for these types of circumstance one may not always choose to subtract the left-hand neighbour of every odd entry, but also choose the right-hand neighbour if subtraction yields the largest detail coefficient in this situation. The largest detail coefficients are generated by using the maximum operator as dual lifting, i.e.,

$$d_{j,k} = s_{j+1,2k+1} - \max(s_{j+1,2k}, s_{j+1,2k+2}). \tag{3.1}$$

For primal lifting we use

$$\begin{aligned} s_{j,k} &= s_{j+1,2k} + \max(0, d_{j,k}, d_{j,k-1}) \tag{3.2}\\ &= s_{j+1,2k} + \max(0, s_{j+1,2k+1} - s_{j+1,2k}, s_{j+1,2k+1} - s_{j+1,2k+2},\\ &\quad s_{j+1,2k-1} - s_{j+1,2k-2}, s_{j+1,2k-1} - s_{j+1,2k}). \tag{3.3} \end{aligned}$$

When comparing primal lifting of this max-lifting scheme with the Haar lifting scheme it may seem a bit peculiar. Detail coefficients are not averaged before adding to $s_{j+1,2k}$. Furthermore, the even entries of s_{j+1} are not updated at all if the "neighbouring" detail coefficients are negative. The choice for this primal lifting may be motivated when regarding some very useful properties of the scheme that are established with this primal lifting operator see [42]. Here, we mention two interesting properties

1. Local maxima of s_{j+1} at even entries are conserved in s_j.
2. Local maxima of s_{j+1} at odd entries are conserved in s_j if no other local maxima appear in a five-entry neighbourhood.

The first property can be verified rather easily. Assuming $s_{j+1,2k}$ is a local maximum, then both $d_{j,k} < 0$ and $d_{j,k-1} < 0$, using Eq. (3.1). Owing to the non-updating for negative $d_{j,k}$ and $d_{j,k-1}$ the local maximum in $s_{j+1,2k}$ is mapped onto $s_{j,k}$ see Equation (3.3). The proof of the second property is left to the reader, as it follows similar arguments.

More general, it can also be shown [42] that max-lifting does not generate new maxima in s_j compared with s_{j+1}. Obviously, s_j is also non-negative if s_{j+1} is a non-negative sequence. By also not averaging in the primal lifting step, max-lifting becomes a procedure that

(a) maps integer-valued s_{j+1} onto integer-valued d_j and s_j,
(b) $0 \leq s_{j+1} \leq M \implies 0 \leq s_j \leq M, |d_j| \leq M$.

Results (a) and (b) make max-lifting an appropriate method for using within, for example, lossless coding on integer-valued data. Furthermore, max-lifting is a very

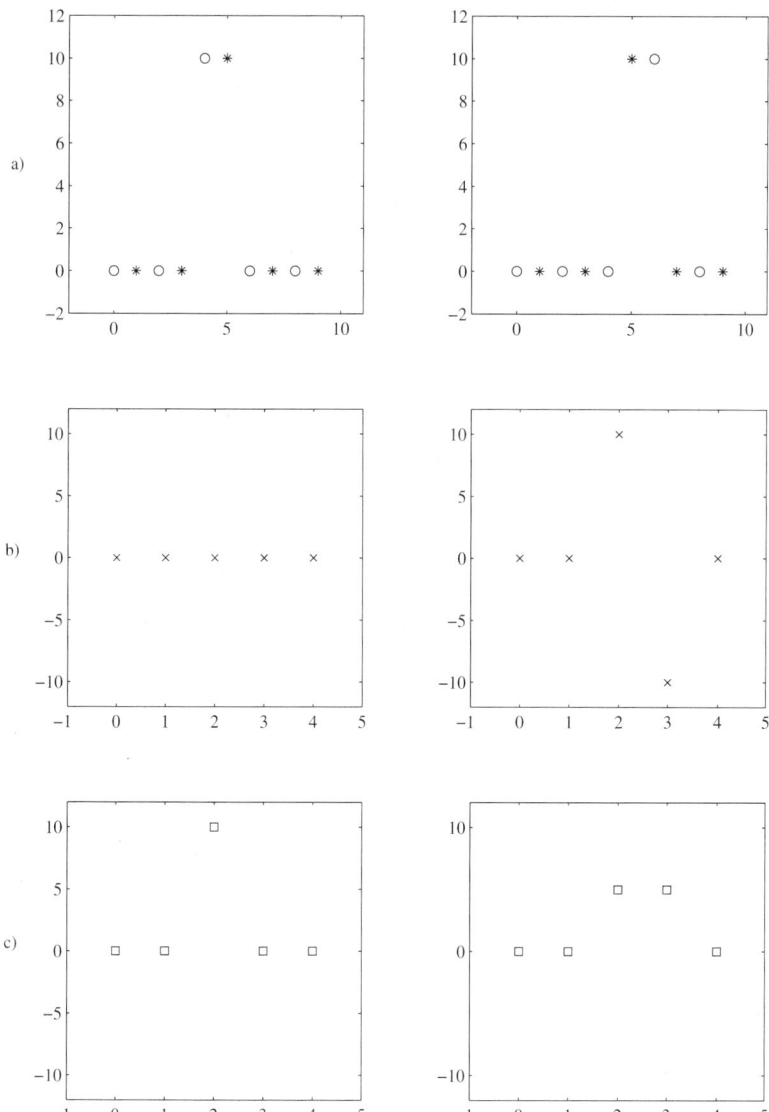

Figure 3.1. Two datasets representing a block function: (a) the original block function (left) and its translate over one unit (right), (b) details d_j of the functions in (a) after prediction with a Haar transform, (c) approximation data s_j of the functions in (a) using primal Haar lifting.

64   3. Nonlinear and Adaptive Lifting

useful tool for image processing, when extended to two dimensions, i.e., taking maxima of nearest neighbours. In grey-value images s_{j+1} each pixel attains a value in the set $\{0, 1, \ldots, 255\}$. After one stage of lifting, again two grey-value images have been created s_j and $|d_j|$.

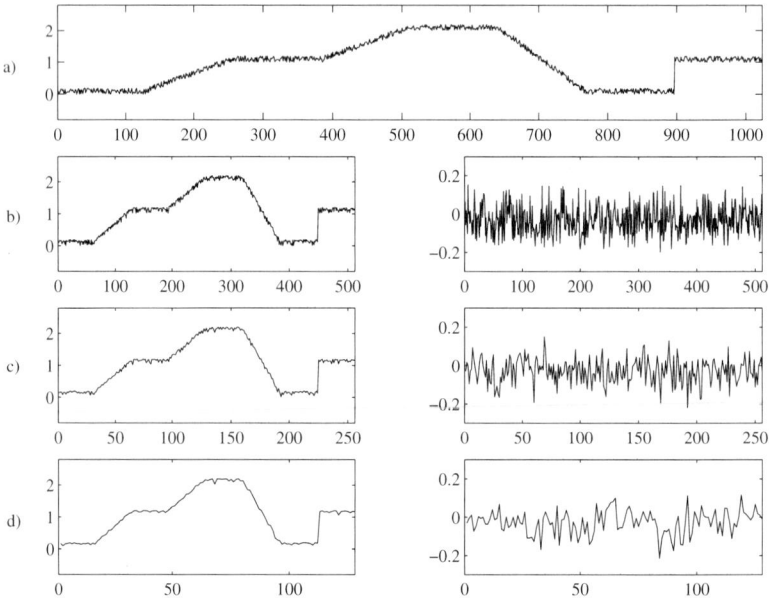

Figure 3.2. The max-lifting scheme with noisy data: (a) a signal consisting of a piecewise second-order polynomial with white noise added to it, (b) approximation data (left) and detail data (right) of the signal in (a) after one stage of the max-lifting scheme, (c,d) as in (b) but after two and three recursive lifting stages respectively.

To illustrate max-lifting we used this scheme on a signal of polynomial order 2 with white noise added to it. This signal has been depicted in Figure 3.2 (upper row). Also the approximation (left) and detail signals (right) generated by applying the scheme three times successively have been visualized. Starting in Figure 3.2.(b) towards the coursest scale in Figure 3.2.(d). Obviously, the max-lifting approach preserved the shape of the original signal at all levels of approximation, while detail coefficients contain white noise components only.

3.1.2 The Median-lifting Scheme

As we have seen, the max-lifting scheme is based on subtracting and adding at each entry only one single element from $s_{j+1,2k}$ to $s_{j+1,2k+1}$ and vice versa. In that way it is closely related to the Haar transform. One can also construct a nonlinear scheme that combines a kind of higher order polynomial interpolation filter with the property that both primal and dual lifting involves exactly one element of $s_{j+1,2k}$ to $s_{j+1,2k+1}$ for computing details $d_{j,k}$ and approximations $s_{j,k}$. A possible way of constructing such a scheme is based on median filtering.

We define the Nth-order median filter of a discrete real-valued signal s as

$$\text{med}_N(s)_k = \text{median}\left(\{s_{k-\lfloor\frac{N}{2}\rfloor}, \ldots, s_{k+\lfloor\frac{N-1}{2}\rfloor}\}\right). \quad (3.4)$$

Observe that the first-order median filter equals the identity operator, since we have $\text{med}_1(s)_k = s_k$. Furthermore, the second-order filter takes the mean of two neighbouring entries, i.e., $\text{med}_2(s)_k = (s_{k-1} + s_k)/2$. Obviously these lowest order median filters coincide with polynomial interpolation filters of the same order.

We start the median lifting scheme by using median filtering in the dual lifting step, yielding

$$d_{j,k} = s_{j+1,2k+1} - \text{med}_N(s_{j+1,2l})_{2k}. \quad (3.5)$$

Here, of course, the median of $s_{j+1,2l}$ and its $N-1$ neighbouring even entries are taken. For primal lifting we use

$$s_{j,k} = s_{j+1,2k} + \frac{1}{2}\text{med}_{\tilde{N}}(d_{j,l})_k, \quad (3.6)$$

with $\tilde{N}$ the order of the median update filter, which may be, but not necessarily, equal to N.

We observe that this median lifting scheme coincides with the Haar lifting scheme as discussed in Section 2.2.1 if we take $N = \tilde{N} = 1$. Other schemes that use median filtering also appear in the literature. In [42] a median lifting scheme was proposed that takes max-lifting as a starting point and replaces the maximum operator by taking the median. Claypoole et al. also discussed the scheme of this section in their paper [10]. Besides this approach, they also suggested an approach where median filtering is used for prediction while polynomial interpolation filters are used for updating the approximation coefficients. In that paper, amongst other things, the use of median-lifting, extended to two dimensions, was investigated for denoising images. Experiments showed that, for images with dominating edges, median lifting for both primal and dual lifting turned out to give better results than when only using median filtering for dual lifting.

As with max-lifting we also used median-lifting on a signal of polynomial order 2 with white noise added to it. Also, median-lifting was applied three times successively and for comparison the computed coefficients were depicted in the same manner as for max-lifting. For this example we have chosen a fourth-order median filter for both prediction and update, i.e., $N = \tilde{N} = 4$. Figure 3.3.(a) shows the noisy signal and its approximation (left) and detail signals (right) at three different levels in Figure 3.3.(b-d). In this example also median-lifting preserved the shape of the original signal at all levels of approximation. The computed detail coefficients contain white noise components as well as a spike induced by the sharp cut-off

66   3. Nonlinear and Adaptive Lifting

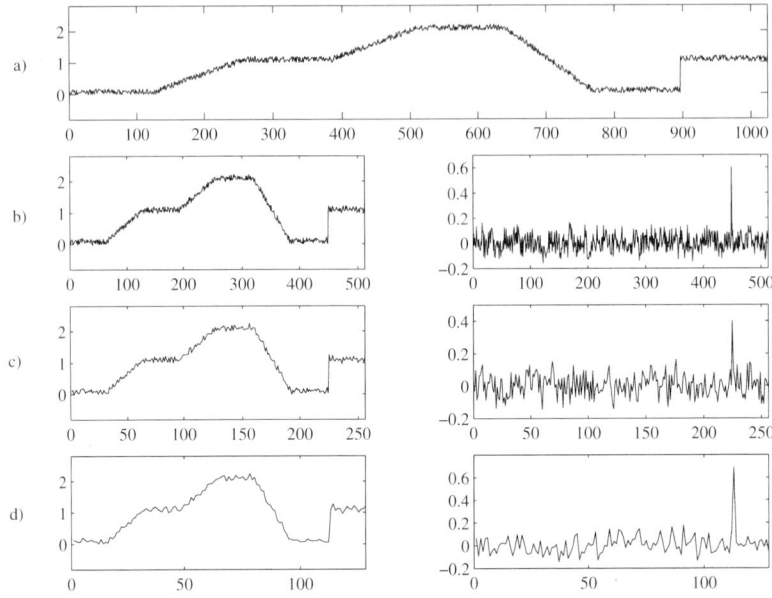

Figure 3.3. The median lifting scheme with noisy data: (a) a signal consisting of a piecewise second-order polynomial with white noise added to it, (b) approximation data (left) and detail data (right) of the signal in (a) after one stage of the median lifting scheme, (c,d) as in (b) but after two and three recursive lifting stages respectively.

(discontinuity) only. We note that such spike may appear as well in the Haar transform, which is, in fact, first-order median filtering. This phenomenon was depicted in Figure 3.1.

3.2 Adaptive Lifting

In the previous section we introduced the concept of nonlinear filtering within the lifting scheme. In particular we discussed max-lifting and median-lifting as examples of nonlinear lifting. Recall the lifting steps in the max-lifting scheme. Dual lifting in this scheme consists of taking the maximum of the nearest even neighbours of each odd entry in s_{j+1} and subtract it from that particular entry. This prediction step can also be regarded in a different way.

We introduce two possible prediction operators, the left Haar predictor or identity operator $P_1(s_{j,2k}) = s_{j,2k}$, and the right Haar predictor or right shift operator $P_2(s_{j,2k}) = s_{j,2k+2}$. Dual lifting in the max-lifting scheme can now be seen as choosing appropriate prediction operators P_1 or P_2. We formalize this approach by introducing a decision operator D^P, based on the even entries $s_{j+1,2k}$, given by

$$D_j^P(k) = \begin{cases} 1, & \text{if } s_{j+1,2k} \geq s_{j+1,2k+2}, \\ 2, & \text{if } s_{j+1,2k} < s_{j+1,2k+2} \end{cases} \quad (3.7)$$

For all pairs of neighbours $s_{j+1,2k}$ and $s_{j+1,2k+2}$ this decision operator attains a value 1 or 2 indicating whether the left or right Haar predictor (P_1 vs. P_2) should be used for dual lifting, i.e.,

$$d_{j,k} = s_{j+1,2k+1} - P_{D^P_{j,k}}(s_{j+1,2k}). \tag{3.8}$$

As for dual lifting, we can also describe the primal lifting step for max-lifting in this way, namely by means of various update operators and a decision operator. As follows from (3.3), this decision operator D^U will only be based on detail coefficients d_j and its outcome can be a choice for the zero operator, the identity operator or the left shift operator. A possible decision map for primal lifting may be

$$D^U_j(k) = \begin{cases} 0, & \text{if } \max(d_{j,k}, d_{j,k-1}) \leq 0, \\ 1, & \text{if } \max(0, d_{j,k-1}) \leq d_{j,k}, \\ 2, & \text{if } \max(0, d_{j,k}) \leq d_{j,k-1}, \end{cases} \tag{3.9}$$

in combination with the set of update operators $U_0 = 0$, $U_1 = 1$ and $U_2 = R$, with R the shift operator on sequences. In fact we have

$$s_{j,k} = s_{j+1,2k} + U_{D^U_{j,k}} d_{j,k}. \tag{3.10}$$

For median-lifting, analoguous decision maps can be defined. However, the number of different cases to be differentiated within the maps rises with the order of the filters N and $\tilde{N}$. The common concept of using "decision maps" for both primal and dual lifting has been generalized to the more general concept of adaptive lifting as depicted in Figure 3.4. There, the lifting scheme has been extended by means of two decision maps, both for dual and for primal lifting. Furthermore, in Figure 3.4, both decision maps are depicted to be dependent on both s_{j+1} and d_j. As we have seen in (3.7) and (3.9), the decisions to be made for picking filters do not need to be induced by the data in both branches of the lifting scheme. Moreover, we will now focus on adaptive lifting schemes in which only prediction or update is based on a decision map, which only depends on the input data for the particular operator. In the next section we will discuss the problem of perfect reconstruction with decisions also dependent on both branches of the scheme.

Of course, adaptive lifting is not just a description of nonlinear lifting schemes, like the max- and median-lifting. This space-adaptive lifting approach was first described by Claypoole et al. [9] for switching between high- and low-order interpolation filters in the lifting scheme. The reason for switching between filters while processing data is that high-order polynomial interpolation filters are desirable for data with low variance, while local irregularities in such data will be smoothed out in the coefficients resulting from the scheme. In the neighbourhood of such irregularities one would like to reduce the order of interpolation such that these significant features are still well captured. An example of such a situation appears in (image) coding. For such application, capturing edges in the lifting coefficients is extremely important, as they determine the contours of an image and, therefore, also to a large extent the whole image. Such irregularities are best reproduced by means of a small

3. Nonlinear and Adaptive Lifting

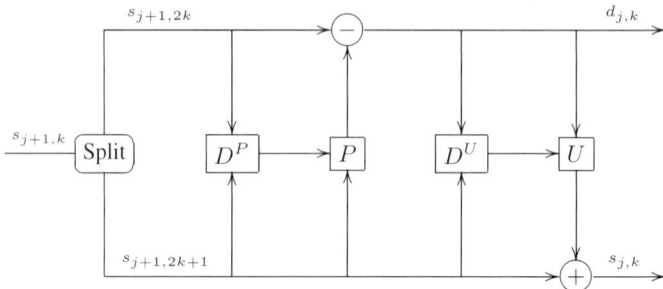

Figure 3.4. A general adaptive lifting scheme: data in both branches of the scheme contribute to the choices of prediction and update filters.

number of coefficients obtained after filtering with a low-order interpolation filter. On the other hand, detailed information of a monotonic background in an image is highly undesirable for image coding. Therefore, it is better to use a high-order interpolation filter when not in the neighbourhood of edges/irregularities and to switch to a low-order interpolation filter near edges.

Switching between filters does not necessarily been a binary choice of prediction filters. In [9] four different prediction filters were used. The update filter was kept fixed at low order. In fact, the Haar update filter, discussed in the previous chapter was used. The prediction filters varied from interpolating filters based on first-, third-, fifth- and seventh-order polynomial interpolation. In fact, the filters were chosen from the Cohen-Daubechies-Feauveau family, CDF N, 1, see [13], with N the order of the interpolating prediction filter and 1 indicating that the Haar update filter is used. So, if we have chosen $N = 1$ the scheme becomes the Haar lifting scheme.

In Figure 3.5 the idea of switching from seventh- to first-order prediction filter has been shown. Depending on the distance towards the edge, the order of the prediction filter decreased and increased again after passing by the edge. The question now is how to design a decision map $D_j^P(k)$. As irregularities can be detected by sudden changes in the derivatives of a function, one might use differences of the $s_{j+1,2k+1}$ with its neighbouring entries and collect them into a gradient vector. Depending on the norm of such a vector we can switch between the order of prediction. However, as is also denoted in Figure 3.5 reduction from seventh- to first-order filter, including fifth- and third-order filters as well, is not possible by considering the differences between one entry and a fixed number of its neighbours.

To be concrete by means of an example we take a data sequence s with a ramp as in Figure 3.5. Starting initially with a 7th order polynomial interpolation filter, we compute the seven-point gradient vector

$$g_7(s)_k = (s_k - s_{k+1}, \ldots, s_k - s_{k+7}).$$

3.2 Adaptive Lifting

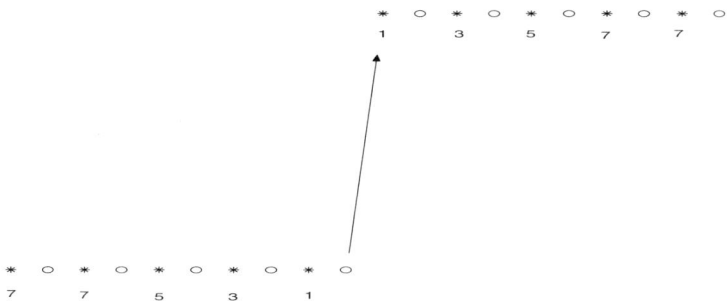

Figure 3.5. Selection of interpolating prediction operators near irregularities.

Then, if $\|g_7(s)_k\|_1 > \Delta_7$, with $\|\cdot\|_1$ the 1-norm of a vector and $\Delta_7 > 0$ a given threshold value, we proceed with a fifth-order polynomial interpolation filter as a predictor and move on to the next sample, otherwise prediction remains seventh-order polynomial interpolation and then move on to the next sample and repeat the procedure at the new position.

When switched to a fifth-order filter one computes the five-point gradient vector $g_5(s)_k = (s_k - s_{k+1}, \ldots, s_k - s_{k+5})$ at the new position. Here three possible actions can be made upon the value of $\|g_5(s)_k\|_1$, namely if $\|g_5(s)_k\|_1 > \Delta_5$, for a given $\Delta_5 > 0$, we proceed with a third-order polynomial interpolation filter as a predictor and move on to the next sample, otherwise we have a closer look at the value of $\|g_5(s)_k\|_1$. If this value becomes too low this can indicate that the ramp in the function is outside the area of interest and so one can switch back to a seventh-order prediction filter, i.e., if $\|g_5(s)_k\|_1 < \Delta_5^*$, for a given $0 < \Delta_5^* < \Delta_5$. If the value of the five-point gradient falls in between these values, the prediction filter remains the same.

The procedure described continues at the next sample with the constraints that one and seven are the lowest and highest orders of filtering in this example. In Figure 3.5 one can see that the gradient vectors attain large values once the ramp becomes part of their domains, and so filter order is decreased. Close to the ramp the lowest order filter is used, while leaving the ramp behind, filter order increases. The numbers below the samples indicate the order of prediction at that particular place.

Instead of a one-sided gradient vector one may also use a double-sided gradient vector and a four-sided gradient vector in two dimensions. In the latter case it is somewhat difficult to indicate a path one should follow for reducing the filter order when starting with order seven. For two dimensions, also possibly for one dimension, we propose a statistical measure on which the decision map D_j^P can be based, namely the so-called relative local variance (RLV). This RLV of a signal w is given by

$$\text{RLV}[w]_k = \sum_{n=k-L}^{k+L} (w_n - \overline{\mu_k})^2 / \text{var}(w), \quad (3.11)$$

with

$$\overline{\mu_{i,j}} = \sum_{n=k-L}^{k+L} w_n/(2L+1)^2, \qquad (3.12)$$

with var(w) the variance of the signal w. The RLV computes locally the variance of the signal and normalizes this value by the total variance. Depending on the value of the RLV one can decide which filter should be used at position k. For the window size L we take the index value of the highest order of prediction filter available, since with this choice all w_n that are used for the prediction of w_k contribute to the RLV. The method of adapting prediction filters based on the RLV is applied on 2-d data in Section 5.1.2.

3.3 Reconstruction of Adaptive Lifting

As we have seen in the previous chapter, when introducing the lifting scheme, perfect reconstruction of s_{j+1} given s_j and d_j is achieved easily by following the forward transform in the reverse direction, as has been depicted in Figure 2.5. Reconstructing s_{j+1} is not obvious anymore for adaptive lifting schemes and, as we will see, an automatic reverse of the scheme is only possible in special cases of adaptive lifting. Of course, choices made in the lifting scheme to obtain s_j and d_j should be followed by analogous choises for primal and dual lifting in the reconstruction scheme.

This means that a general adaptive scheme as in Figure 3.4 should be followed by a general adaptive reconstruction scheme, as has been visualized in Figure 3.6 involving decision maps $\widetilde{D}_j^U$ and $\widetilde{D}_j^P$. These decision maps will not equal D_j^U and D_j^P in the most general case, which is also suggested by Table 3.1. Here, the input sequences for the four separate decision maps are written down. We observe that since each decision map depends on different input data, inversion by means of equal maps can only be achieved in some special cases of the general adaptive scheme of Figure 3.4. Moreover, it is easily shown, that for certain classes of adaptive schemes, inversion can only be guaranteed by keeping track of the choices made for prediction and/or update filters. If this is the case it is a main disadvantage for such schemes when used for applications where a sender/receiver channel is built in between decomposition and reconstruction, e.g., in coding. In the latter application, besides the coded s_j and d_j a book-keeping file also has to be transmitted to the decoder to reconstruct the data into s_{j+1}.

3.3.1 Automatic Perfect Reconstruction

When not following the most general adaptive scheme as in Figure 3.4, perfect reconstruction may become as trivial as in the non-adaptive case. Table 3.1 shows that both D_j^P and $\widetilde{D}_j^P$ depend on $s_{j+1,2k}$. If these data are also the only data the decision maps depend on, $\widetilde{D}_j^P$ coincides with D_j^P. The same argument holds for the

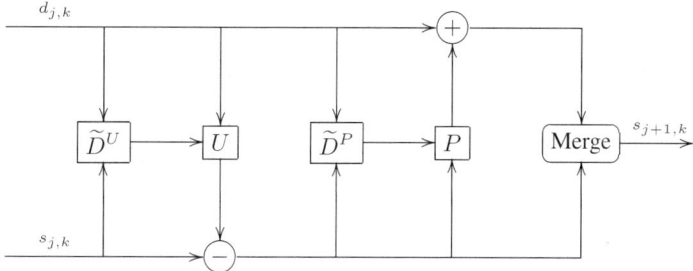

Figure 3.6. Reconstruction with a general adaptive lifting scheme: as in the general adaptive decomposition scheme, data in both branches of the scheme contribute to the choices of prediction and update filters.

update decision maps D_j^U and $\widetilde{D}_j^U$ depending on $d_{j,k}$. This means that a sufficient condition for perfect reconstruction, without keeping track of the decision we make, is that D_j^P should depend only on $s_{j+1,2k}$ and D_j^U only on $d_{j,k}$. Such an adaptive lifting scheme is represented by Figure 3.7.

Table 3.1. The decision maps involved in a perfect reconstruction lifting scheme and their possible input sequences.

Decision map	Input data
D_j^P	$s_{j+1,2k}, s_{j+1,2k+1}$
$\widetilde{D}_j^P$	$s_{j+1,2k}, d_{j,k}$
D_j^U	$s_{j+1,2k}, d_{j,k}$
$\widetilde{D}_j^U$	$s_{j,k}, d_{j,k}$

Of course, using only a decision map D_j^P to adapt dual lifting combined with fixed primal lifting can also be reconstructed in the sammer manner, in the case, that D_j^P only depends on $s_{j+1,2k}$. Decision map D_j^U is then made constant, independent of the input data $d_{j,k}$ and $s_{j+1,2k}$. On the other hand, using only a decision map D_j^U to adapt primal lifting can be reconstructed in the same manner by making D_j^P constant, independent of the input data $s_{j+1,2k+1}$ and $s_{j+1,2k}$.

As denoted before, an interesting topic is how to built lifting schemes where equivalent decision maps appear in analyzing and reconstructing, where it is not necessary to keep track of the decisions be made and which are based on data in both branches of the lifting scheme. Using an example of such a situation will show that building such schemes is highly non-trivial, although it is highly desirable to have such schemes.

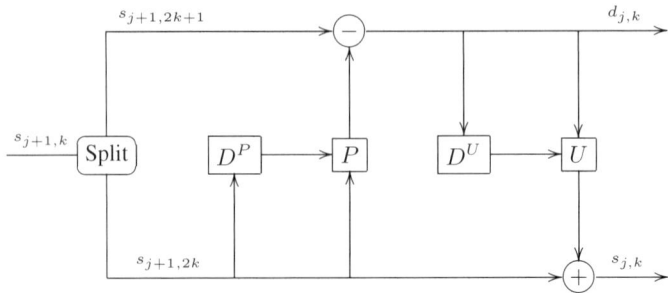

Figure 3.7. Perfect reconstruction without book-keeping is easily achived by letting the decision maps only depend on the input data of the corresponding prediction and update filters.

In the example here we concentrate on adaptive primal lifting for simplicity reasons. Furthermore, as the derivations become rather technical, we introduce the abbreviations $v_k = s_{j+1,2k}$, $\tilde{v}_k = s_{j,k}$ and $w_k = d_{j,k}$ for fixed scale j. The decision map we use is based on the sequence of gradient vectors $g(v, w)$ defined as

$$g(v, w)_k = (v_k - w_{k-1}, v_k - w_k). \qquad (3.13)$$

High values of the gradient vector can be associated with high-frequency components, while low values are appropriate for a low-frequency content locally. As high valued gradient vectors often appear near edges, adaptive lifting becomes an interesting tool for switching between a smoothing update filter and a high-frequency preserving filter. Here we suggest the decision map

$$D^U[v, w](k) = \begin{cases} 1, & \text{if } \|g(v, w)_k\|_1 \leq \Delta, \\ 2, & \text{if } \|g(v, w)_k\|_1 > \Delta, \end{cases} \qquad (3.14)$$

with $\Delta > 0$ a threshold value to be chosen upon the data and the application one has in mind. Two possible update operators are related to the decision map:

$$U_i w_k = u_{i,1} w_{k-1} + u_{i,2} w_k, \quad i = 1, 2, \qquad (3.15)$$

involving filter coefficients $u_{1,1}, u_{1,2}$ for $D^U[v, w]_k = 1$ and $u_{2,1}, u_{2,2}$ in the case that $D^U[v, w]_k = 2$. Observe that the update filter uses exactly the same data of w as the decision map does. In a later stage we will generalize this concept. In addition to classical lifting we also rescale v_k by means of a scaling coefficient $a_1, a_2 > 0$ depending on the value of $D^U[v, w]_k$ as well. A rescaling procedure like this was also used for the unbalanced Haar transform in the previous chapter. Note that if $a_i = 1$ rescaling cancels out of the scheme. Resuming, we get approximation coefficients

$$\tilde{v}_k = a_i v_k + u_{i,1} w_{k-1} + u_{i,2} w_k, \quad i = 1, 2, \qquad (3.16)$$

depending on the value of $i = D^U[v, w]_k$. This concept is depicted in Figure 3.8.

3.3 Reconstruction of Adaptive Lifting

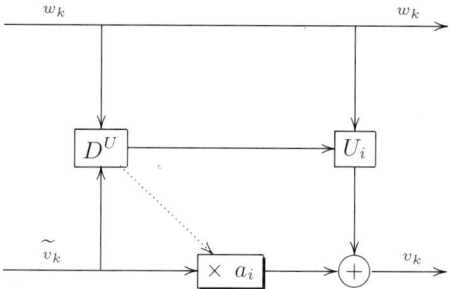

Figure 3.8. Additional scaling of the data is needed to guarantee perfect reconstruction if the decision map depends on the data in both channels.

Now that the decision map is based on both v and w, the problem we have to deal with, whether v can be recovered from $\tilde{v}$ without transmitting the subsequent values for $D^U[v, w]_k$. From (3.16) the solution for this problem seems trivial, as this equation can be rewritten as the inversion formula

$$v_k = (\tilde{v}_k - u_{i,1} w_{k-1} + u_{i,2} w_k)/a_i, \quad i = 1, 2. \tag{3.17}$$

However, it is far from obvious how to get the knowledge of using which coefficient a_i to involve in rescaling and which $u_{i,1}$ and $u_{i,2}$ to use as filter coefficients for updating at the reconstruction side. A similar decision map $\tilde{D}^U[\tilde{v}, w]_k$ has to answer these questions. So the problem arises for which $\tilde{v}$ we have

$$D^U[v, w] = \tilde{D}^U[\tilde{v}, w],$$

with $\tilde{D}^U$ be given by

$$\tilde{D}^U[\tilde{v}, w](k) = \begin{cases} 1, & \text{if } \|g(\tilde{v}, w)_k\|_1 \leq \tilde{\Delta}, \\ 2, & \text{if } \|g(\tilde{v}, w)_k\|_1 > \tilde{\Delta}, \end{cases} \tag{3.18}$$

with $\tilde{\Delta}$ an appropriate threshold value for the reconstruction part. This situation is illustrated in Figure 3.9.

Since $\tilde{v}$ depends on the coefficients a_i, $u_{i,1}$ and $u_{i,2}$, $i = 1, 2$, the problem we are dealing with can be translated into the question of chosing these six parameters in an appropriate way. The remainder of this section is devoted to deriving necessary and sufficient conditions on the coefficients such that automatic perfect reconstruction is possible for this particular example.

A first surprising result is that perfect reconstruction in this case is only possible if

$$a_1 + u_{1,1} + u_{1,2} = a_2 + u_{2,1} + u_{2,2}. \tag{3.19}$$

By defining $S_i = a_i + u_{i,1} + u_{i,2}$ we can translate this condition into $S_1 = S_2 = 1$ after normalization. For proving this result we assume that $S_1 \neq S_2$ and that perfect

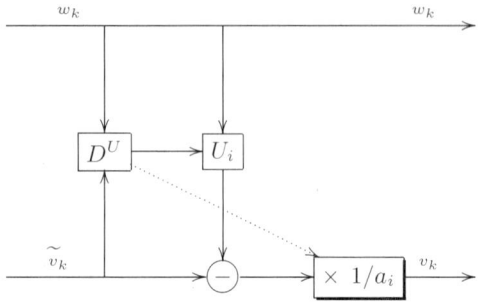

Figure 3.9. Before merging data together yielding the original input signal, scaling of the data is needed to guarantee perfect reconstruction. The scaling parameter is given by the outcome of the decision map.

reconstruction is achieved. Next we consider two different cases involving signals v_1, w_1 and v_2, w_2, with $v_1 \neq v_2$ and $w_1 = w_2 = w$. Since perfect reconstruction is assumed we should get $\tilde{v}_1 \neq \tilde{v}_2$. We show by construction that for different v_1 and v_2 it is possible to end up with $\tilde{v}_1 = \tilde{v}_2$, yielding the rejection of $S_1 \neq S_2$.

For both situations we take $w = c$, a sequence of a certain constant $c > 0$. For v_1 we take $v_1 = c + r_1$ with r_1 a sequence satisfying $|r_{1,k}| \leq \Delta/2$. This means that
$$\|g(v_1, w)_k\|_1 = |v_{1,k} - w_{k-1}| + |v_{1,k} - w_k| = 2|r_{1,k}| \leq \Delta,$$
yielding $D^U[v_1, w] = 1$. After updating we get, using (3.16),
$$\begin{aligned}\tilde{v}_{1,k} &= a_1 v_{1,k} + u_{1,1} w_{k-1} + u_{1,2} w_k \\ &= a_1 r_{1,k} + a_1 c + u_{1,1} c + u_{1,2} c = a_1 r_{1,k} + c S_1 \end{aligned} \quad (3.20)$$

For v_2 we take $v_{2,k} = c + (c(S_1 - S_2) + a_1 r_{1,k})/a_2$. Furthermore, we assume at this stage that the constant c satisfies
$$c > \frac{\Delta(a_1 + a_2)}{2|S_1 - S_2|}.$$

Observe that since we assumed $S_1 \neq S_2$, choosing such constant c and constructing data v_2 in this manner is possible. With these choices we get
$$\begin{aligned}\|g(v_2, w)\|_1 &= 2|v_{2,k} - c| \geq 2c|(S_1 - S_2)|/a_2 - a_1|r_{1,k}|/a_2 \\ &> \frac{\Delta(a_1 + a_2)}{a_2} - \frac{a_1 \Delta}{a_2} = \Delta,\end{aligned}$$

yielding $D^U[v_2, w] = 2$. For this situation after updating, as in (3.16), we get
$$\begin{aligned}\tilde{v}_{2,k} &= a_2 v_{2,k} + u_{2,1} w_{k-1} + u_{2,2} w_k \\ &= c(S_1 - S_2) + a_1 r_{1,k} + a_2 c + u_{2,1} c + u_{2,2} c \\ &= c(S_1 - S_2) + a_1 r_{1,k} + c S_2 = a_1 r_{1,k} + c S_1.\end{aligned} \quad (3.21)$$

3.3 Reconstruction of Adaptive Lifting   75

Comparing (3.20) and (3.21) we see that primal lifting of two different input datasets may result in the same updated signal $\tilde{v}_1 = \tilde{v}_2$. Using inversion formula (3.17) it turns out that $v_1 = v_2$, which is not the case. This contradiction results in the rejection $S_1 \neq S_2$. So, from now on we consider normalized coefficients that satisfy

$$a_i + u_{i,1} + u_{i,2} = 1, \quad i = 1, 2, \tag{3.22}$$

Having derived this necessary relation between the coefficients we return to the problem, whether $D^U[v, w] = \tilde{D}^U[\tilde{v}, w]$. The answer to this problem is given by comparing the gradient vectors $g(\tilde{v}, w)_k$ and $g(v, w)_k$. Using (3.16) and (3.22) we can write

$$\begin{aligned}
\tilde{v}_k - w_{k-1} &= a_i v_k + (u_{i,1} - 1) w_{k-1} + u_{i,2} w_k \\
&= (1 - u_{i,1} - u_{i,2}) v_k + (u_{i,1} - 1) w_{k-1} + u_{i,2} w_k \\
&= (1 - u_{i,1})(v_k - w_{k-1}) - u_{i,2}(v_k - w_k).
\end{aligned}$$

Similarly, we can also derive

$$\tilde{v}_k - w_k = -u_{i,1}(v_k - w_{k-1}) - (1 - u_{i,1})(v_k - w_k). \tag{3.23}$$

Combining these two equations yields the following relation for the gradient vectors:

$$g(\tilde{v}, w) = M_i\, g(v, w), \quad i = D^U[v, w], \tag{3.24}$$

with M_i a 2×2 matrix given by

$$M_i = \begin{pmatrix} 1 - u_{i,1} & -u_{i,2} \\ -u_{i,1} & 1 - u_{i,2} \end{pmatrix}. \tag{3.25}$$

Note that $\det(M_i) = 1 - u_{i,1} - u_{i,2} = a_i$ due to (3.22). This relation holds the solution to the problem of perfect reconstruction in this example, since the condition $D^U[v, w] = \tilde{D}^U[\tilde{v}, w]$ is satisfied if given a threshold value $\Delta > 0$ there exist a $\tilde{\Delta} > 0$ such that

$$\begin{aligned}
(\|g(v, w)\|_1 \leq \Delta) &\implies (\|M_1\, g(v, w)\|_1 \leq \tilde{\Delta}), \\
(\|g(v, w)\|_1 > \Delta) &\implies (\|M_2\, g(v, w)\|_1 > \tilde{\Delta}).
\end{aligned}$$

It can be shown rather easily that this is the case if we have

$$\|M_1\|_1 \leq \|M_2^{-1}\|_1^{-1}. \tag{3.26}$$

Obviously, if $\|g(v, w)\|_1 \leq \Delta$ and we take $\tilde{\Delta} = \|M_1\|_1 \Delta$, then

$$\|M_1 g(v, w)\|_1 \leq \|M_1\|_1 \cdot \|g(v, w)\|_1 \leq \|M_1\|_1 \Delta = \tilde{\Delta}.$$

On the other hand, if $\|g(v, w)\|_1 > \Delta$, then

$$\tilde{\Delta} = \|M_1\|_1 \Delta < \|M_1\|_1 \cdot \|g(v, w)\|_1 \leq \|M_2^{-1}\|_1^{-1} \cdot \|g(v, w)\|_1 < \|M_2\, g(v, w)\|_1.$$

76   3. Nonlinear and Adaptive Lifting

The latter inequality follows from

$$\|g(\boldsymbol{v},\boldsymbol{w})\|_1 = \|M_2^{-1}M_2 g(\boldsymbol{v},\boldsymbol{w})\|_1 \leq \|M_2^{-1}\|_1 \cdot \|M_2 \, g(\boldsymbol{v},\boldsymbol{w})\|_1.$$

Now that we have found a condition on the coefficients to get perfect reconstruction, the question arises, to whether we can find such coefficients. A very simple case for which perfect reconstruction holds is if we take

$$0 \leq u_{1,1}, u_{1,2} \quad \text{and} \quad u_{2,1}, u_{2,2} \leq 0, \tag{3.27}$$

with the constraint that $u_{1,1} + u_{1,2} = 1 - a_1 < 1$. In this particular situation we have $\|M_1\|_1 = 1$ and $\|M_2^{-1}\| \leq 1/a_2 \leq 1$, as M_2^{-1} is given by

$$M_2^{-1} = \frac{1}{a_2} \begin{pmatrix} 1 - u_{2,2} & u_{2,2} \\ u_{2,1} & 1 - u_{2,1} \end{pmatrix}. \tag{3.28}$$

The example we have discussed shows that it is rather complicated to guarantee perfect reconstruction in an adaptive lifting scheme where decision maps depend not just on the data the prediction or update operator is acting on. However, we have also shown that such schemes exist. Moreover, in [65] it was shown that perfect reconstruction can also be achieved by extending the gradient vector and corresponding update filter to an arbitrarily large number of neighboring entries, i.e., with gradient vector

$$g(\boldsymbol{v},\boldsymbol{w}) = (v_k - w_{k-L_1}, \ldots, v_k - w_k, \ldots, v_k - w_{k+L_2}), \quad L_1, L_2 \geq 0 \tag{3.29}$$

and two possible update operators, related to the decision map,

$$U_i w_k = \sum_{n=0}^{L_1+L_2} u_{i,n+1} w_{k-L_1+n}. \quad i = 1, 2. \tag{3.30}$$

A study by Piella [64] discussed the problem of perfect reconstruction in a general fashion. Several classes of decision maps and update filters were discussed extensively. Amongst others was the example we discussed in this section.

4. Numerical Condition

This chapter is probably the most mathematical one in this book. To some readers, with a more application-oriented background, the discussion in this chapter might look a bit theoretical. Yet, the practical relevance of this theoretical analysis should not be underestimated. In spite of the possible impression the reader gets from studying this chapter, the theory in these sections is kept at a very moderate level. We hope that, for those readers confronted with stability issues in their applications, the text presented may serve as an introduction for the exploration of the rich, yet quite heavy, literature on the subject.

Numerical condition is an important underlying issue in the construction of second-generation wavelets. Classical wavelet theory starts from the concept of a multiresolution analysis. This concept imposes a stable scaling basis at a given scale, and in the classical, regular setting it implies the existence of a stable wavelet basis in a multiscale decomposition. The conditions necessary for the existence of such a stable scaling basis coincide with the conditions for convergence of the subdivision scheme [12, 73, Section 7.2]. In other words, if the basis is not stable, then the mother wavelet or father scaling function cannot be found numerically by running a subdivision scheme. If subdivision does not converge, then the two-scale filters are not useful in practical applications. Hence, all classical wavelet transforms used in practice are stable. Since a classical wavelet decomposition is a sequence of identical filter operations, the conditions for stability and subdivision convergence can be formulated in terms of eigenvalues of one single matrix describing the transition from one scale to the next one. Such a simple criterion is obviously no longer possible in the irregular setting, where the filterbank coefficients are scale dependent. Since the basis is no longer a dilation and translation of a single mother function, there is no immediate link between convergence of a subdivision scheme for such a function and stability. The impact of the irregularity in the data locations is, therefore, the central issue in this chapter.

With respect to this irregularity, we can ask two questions. The first concerns how to construct stable refinements from a give finite resolution. Second, we can investigate how to construct stable multiscale decompositions for data on a *given* grid. There are two possible directions to follow here: one is to concentrate on the construction of stable decompostions by lifting *existing* stable decompositions; the second possibility is the construction of decompositions from the very beginning. While the first possibility concentrates on primal and dual lifting steps, the second one may also consider the splitting scheme, as we discuss below. This chapter is mainly concerned with the construction of stable decompositions on equidistant grids. Applications of both questions are, of course, different. Whereas the first question shows up in domains like computer graphics, the second one is related to problems in the analysis of data on given locations.

We first start with an example that illustrates the importance of stability in a practical application. Next, in Section 4.2, we define some key concepts in stability and explain that the description of wavelet basis stability requires a treatment in infinite dimensional function spaces, even if we work with a finite number of samples in practice. Section 4.3 establishes general criteria for stable decompositions and concentrates on criteria specific for multiresolution analyses. One of the necessary (but not sufficient) conditions turns out to be that the wavelets have at least one primal vanishing moment; see Section 4.4.2. Section 4.4 concentrates on the design of lifting steps for stable decompositions.

4.1 Stability in Wavelet Smoothing on Irregular Point Sets

As in Chapter 2, we start the discussion with an illustration from data smoothing by wavelet thresholding. Suppose we are given a set of N irregularly spaced observations y at locations x. The locations x_i are fixed, but can be thought of as originally taken from a uniform random distribution on the interval $[0, 1]$. At first sight, this grid, though irregular, looks globally balanced, i.e., there are no regions with sparse observations together with densely sampled intervals. The observations themselves are subject to noise $y_i = f(x_i) + \varepsilon_i$, where we suppose that the noise is homoscedastic (i.e. with constant variance), uncorrelated (i.e. white) and the underlying function f is piecewise smooth.

As an example, we take a noisy version of the heavysine test function [39] evaluated in $N = 2048$ points, uniformly distributed on $[0, 1]$; see Figure 4.1. We apply

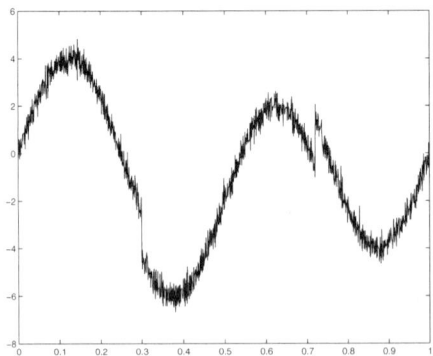

Figure 4.1. $N = 2048$ noisy observations from the heavisine test function.

4.1 Stability in Wavelet Smoothing on Irregular Point Sets

a second-generation wavelet transform to these observations, with in each step an even-odd splitting, followed by a cubic polynomial prediction and a two-taps update. (For those readers not familiar with signal processing terminology, this two-taps update corresponds to a band-limited U-matrix with two nonzero diagonals.) The two degrees of freedom of the update step are used to obtain primal wavelets with two vanishing moments. We also compute the corresponding noise-free wavelet coefficients. In order to remove the noise, we apply a (soft) threshold to the coefficients, after we normalized them w.r.t. the variance. Indeed, since the transform is not orthogonal, the wavelet coefficients do not have constant variance (i.e., noise level), even if the input was stationary noise, i.e., noise with constant variance (often called homoscedastic noise in statistical literature). Fortunately, it is easy and fast to normalize the coefficients such that they do have constant variance [46, Chapter 7]. Since we also know the noise-free coefficients in this simulation, we can assess the threshold that minimizes the mean-square error of the thresholded coefficients compared with the noise-free values. (In practical situations, this threshold has to be estimated, e.g., using (generalized) cross-validation [46] or SURE [39].) The output of this thresholding procedure is depicted in Figure 4.2(a). Surprisingly, this result shows some spurious, unacceptable effects. At first sight, it looks as if some coefficients have been blown up, rather than thresholded: the artifacts have indeed the shape of one or two wavelet basis functions. The problem must be explained, at least partly, by the irregularity of the grid, as illustrated by the output in Figure 4.2(b). This function estimation was obtained by running exactly the same transform, using the lifting scheme and identical options for transform depth, splitting, prediction, update and thresholding policy, except that we pretended as if the data were on an equidistant grid. In other words, the data were mapped onto this equidistant grid, without any further preprocessing, and are remapped onto the true, non-equidistant grid after smoothing. Obviously, the output reflects the irregularity of the grid, but at least it does not show the deformations of the second-generation transform.

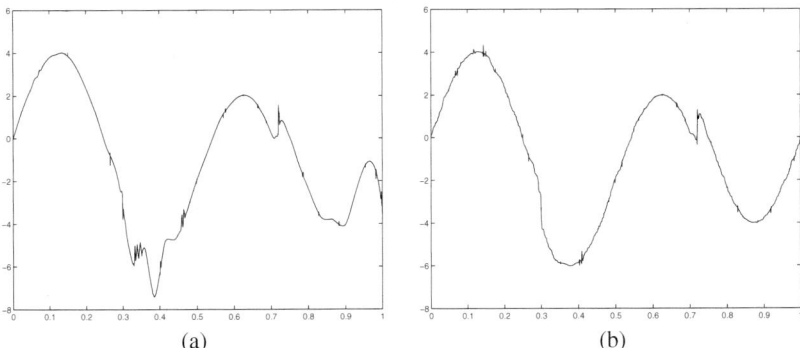

Figure 4.2. Output of classical wavelet thresholding applied to second-generation wavelet coefficients of data in Figure 4.1. This output shows some spurious, unacceptable effects.

4. Numerical Condition

The problem lies in the fact that the non-equispaced transform with cubic prediction and two-taps update is unstable, i.e., far from orthogonal. The instability is further attenuated by the variance normalization step. Altogether, it may happen that small coefficients in the normalized representation turn out to carry crucial information, while others are large but mainly noise. Indeed, since the basis is arbitrarily oblique, some noisy, non-smooth features do not fit well into this basis: a small noisy structure may require a couple of large, mutually annihilating coefficients. A geometric example in $\mathbb{R}^3$, depicted in Figure 4.3, illustrates what is going on. Suppose we have the basis vectors $\{(-1/2, \sqrt{3}/2, 0), (-1/2, -\sqrt{3}/2, 0), (1, 0, \varepsilon)\}$ and the noise vector is $(0, 0, \varepsilon)$ in the canonical basis. Its coordinates in this oblique basis are $(1, 1, 1)$. If one or two of these coordinates are thresholded, then "hidden components" become visible, as if a coefficient had been blown up rather than thresholded. This bad condition can only be detected with a global analysis: none of the basis vectors is close to another one. If the wavelet transform is far from

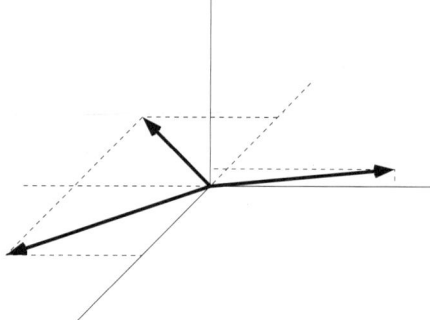

Figure 4.3. An arbitrarily unstable basis in $\mathbb{R}^3$.

orthogonal, then it is hard to predict what effect an operation in the wavelet domain is going to have after reconstruction. As a consequence, the behaviour of the mean-square error in terms of wavelet coefficients can be quite different from the error after reconstruction. Finding a proper threshold value from wavelet coefficients only is hard, if at all possible; in many situations, even the smallest threshold value induces unacceptable side-effects.

One approach [46, Chapter 7] could be to map and remap from an equispaced grid, as in Figure 4.2(b), and use this estimation as a *prototype*, or pilot estimator. In order to exploit the grid-adaptivity of a second-generation approach to its full extent, we compute the second-generation wavelet coefficients of this pilot estimator and threshold them whenever it is safe to do so. Whether it is safe to threshold a given coefficient can be verified by comparing the pilot estimator at the corresponding location with the (unstable) reconstruction from a second-generation approach at that location.

Although this works fine in most cases, it does not fully exploit the possibilities of a smooth, grid-adaptive second-generation approach. Secondly, the result is limited by the quality of the pilot estimator. Figure 4.4 sketches a situation where mapping onto equispaced grids may result in a loss of information. The figure shows a function with a long tail on the right-hand side, along with 128 noisy observations on a non-equidistant point set. The grid is a smooth, exponential function of the equidistant grid, so smoothness is not the issue in this example. The observations of the tail are, however, sparse. As the dashed line illustrates, a significant deviation from the true function is hard to distinguish after mapping the data onto the equidistant grid. It is clear that the balance between closeness of fit and smoothness should depend on the observation density. A wavelet transform constructed on an equidistant grid pays too little attention (weights) to fitting the data on the right-hand side.

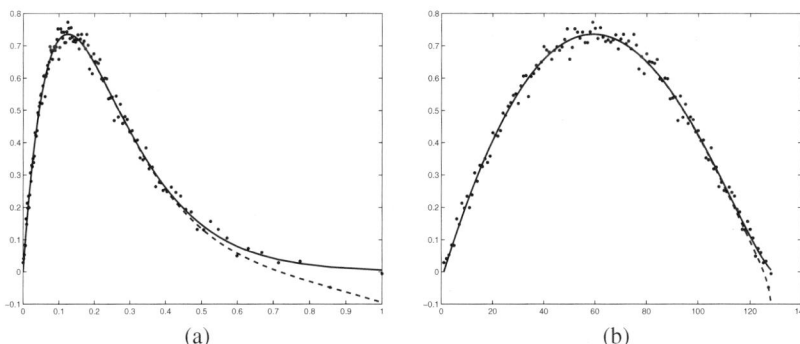

Figure 4.4. An example of a dataset where an important feature, the tail on the right-hand side, is almost lost by mapping the data onto a regular grid.

Therefore, we start from a different viewpoint in this chapter. We analyze the origin of instabilities in second-generation wavelets and look for more stable implementations. This means that the focus on vanishing moments has to be shifted towards other criteria. Yet, we try to extend the decompositions in such a way that they coincide as much as possible with the classical implementation when applied to equidistant grids.

4.2 Condition from Finite to Infinite Dimensions

4.2.1 Condition Numbers

Definition 4.1 *The condition number of a linear transform with matrix $A \in \mathbb{R}^{n \times n}$ is defined as*

$$\kappa(A) = \|A\|_2 \cdot \|A^{-1}\|_2, \tag{4.1}$$

where $\|A\|$ is the 2-norm of a matrix A:

$$\|A\|_2 = \max_{x \neq 0} \frac{\|Ax\|_2}{\|x\|_2}. \tag{4.2}$$

Given this definition of a matrix (operator) norm, matrices with finite norm are sometimes called *bounded*, and matrices with finite condition number are called *bounded and boundedly invertible*. This condition number also equals the ratio of the largest and smallest singular value of A. If A is singular (non-invertible), the smallest singular value is zero, and $\kappa(A) = \infty$. If A is non-singular, then the columns ψ_i of its inverse serve as a basis for $\mathbb{R}^n$, i.e., there is no non-trivial way to decompose $\mathbf{0}$ as a linear combination of ψ_i, and, on the other hand, every vector $x \in \mathbb{R}^n$ can be written as a linear combination of ψ_i:

$$x = \sum_{i=1}^n w_i \psi_i,$$

or, equivalently, since $A^{-1} = [\ldots \psi_i \ldots]$:

$$x = A^{-1} \cdot w,$$

where the coefficients w are found by the *forward* transform

$$w = A \cdot x.$$

Since every lifting step is immediately invertible, the inverse of a wavelet transform constructed by successive lifting steps trivially exists. As long as we are working on a finite resolution, the condition number of the wavelet transform is finite. This formalism is not sufficient to describe the instabilities observed. Indeed, as the resolution level can grow arbitrarily fine, the condition number can become arbitrarily large.

4.2.2 Stable Bases

We start working in an (infinite dimensional) Banach space $\mathcal{B}$, i.e., a vector space which has a norm and which is complete under the metric induced by that norm. Typical examples are the L_p function spaces. In such an inifite dimensional space, we make a distinction between an *algebraic* or *Hamel* basis and a *countable* or *Schauder* basis. The first one is the classical notion of a basis, in that it requires that every element of the vector space can be uniquely decomposed as a *finite* linear combination of the basis vectors. The existence of such a linearly independent spanning set is equivalent to the Axiom of Choice. However, we follow the second direction, that of countable bases. A Schauder basis or countable basis is a countable set of vectors $\psi_i, i \in \mathbb{N}$ such that every element f of the Banach space can be written as an *expansion*:

$$f = \sum_{i=1}^\infty a_i \psi_i, \tag{4.3}$$

where the coefficients a_i are unique and the equality should be read as convergence of the right-hand side to the left-hand side. Wavelet transforms, as well as Fourier expansions (of periodic functions), are typical examples of Schauder basis decompositions. The requirement of convergence for any function f is equivalent to saying that any f can be approximated arbitrarily well by a finite linear combination of the basis functions. In other words, the subspace defined by finite linear combinations of the Schauder basis is dense; and since it is also countable, the Banach space that we are working in must be separable, i.e., it has a countable, dense subspace.

If the norm is induced by a scalar product, then the Banach space we are working in is unitary, which makes it a Hilbert space. In such a space, we can define orthogonality (two vectors are orthogonal if their scalar product equals zero), and a countable basis is called a total *orthogonal system* if all basis vectors (basis functions) are mutually orthogonal. Orthogonality is a very interesting property, since coefficients a_i of convergent decompositions satisfy Parseval's (or Plancherel's) equality:

$$\|f\|^2 = \sum_{i \in \mathbb{N}} |a_i|^2 \|\psi_i\|^2.$$

In other words, the ℓ_2-norm of the sequence a of coefficients is equivalent to the norm of f. Since this sequence norm depends on coefficient magnitudes only (and not on signs or phases), magnitudes of (a countable set of) coefficients *unconditionally* determine whether or not f belongs to the Hilbert space. This notion of *unconditional basis* can be generalized for not necessarily unitary Banach spaces.

Example. Legendre polynomials $P_n(x)$ are an orthogonal basis for functions in $L_2([-1, 1])$. Every function in this Hilbert space has a unique, convergent decomposition

$$f(x) = \sum_{k=0}^{\infty} a_k P_k(x).$$

Legendre polynomials in their turn can be written as a linear combination of the form

$$P_n(x) = \sum_{m=0}^{n} b_{n,m} x^m,$$

so $x^m, m \in \mathbb{N}$ is equally a Schauder basis of $L_2([-1, 1])$. The proposed power series is possible for all L_2-functions on $[-1, 1]$, so not only for functions with Taylor expansion, and the proposed construction does not (necessarily) coincide with the Taylor expansion. In spite of this unique decomposition for every L_2-function, the basis x^m does not meet what we require from a 'good' (i.e., unconditional) basis. Indeed, a basis decomposition is typically performed to characterize, analyze or even manipulate the function by its coefficients. Consider the example

$$f(x) = \frac{1}{1 + x^2}.$$

This L_2-function has a Taylor expansion

$$f(x) = \sum_{k=0}^{\infty} (-1)^k x^{2k}.$$

Since the n-term approximation by this expansion converges quadratically to $f(x)$, the above equality also holds in L_2-sense; this is the unique decomposition into the Schauder basis we are looking for. If we replace all coefficients in this expansion by their absolute values, we get

$$\sum_{k=0}^{\infty} x^{2k} \sim \frac{1}{1-x^2}.$$

The resulting function is no longer in $L_2([-1,1])$! Nevertheless, switching signs is a relatively small operation. Wavelet threshold algorithms, for instance, would select exactly the same set of important coefficients in both decompositions. Yet, this small operation leads to a fundamentally different conclusion. Although $f(x) \in L_2([-1,1])$, this conclusion cannot be made from the magnitude of its coefficients in the x^m-expansion. The contributions of components x^m compensate for each other. We could put the contributions in a different order; for instance, we could start with n positive coefficients, then a single negative, then $n+1$ positives and so on. The negatives would then come too late to compensate for the positives and the series would diverge, simply by reordering! The fact that one contribution partly undoes the effect of another is, of course, due to the skewness of the basis. Although we do not necessarily want orthogonality, we do want to exclude bases where this phenomenon of compensation leads to essentially different conclusions. Indeed, we would like to draw conclusions from individual contributions separately, without conditioning on the entire set of coefficients.

Definition 4.2 $\{\psi_i\}, i \in \mathbb{N}$ *is an unconditional basis of a Banach space $\mathcal{B}$ (typically $L_p[0,1]$) if $\sum_i |w_i|\psi_i(x) \in \mathcal{B}$ for all $\sum_i w_i \psi_i(x) \in \mathcal{B}$ and vice versa.*

This means that we can define a norm for the coefficient vector $\boldsymbol{w}$ as

$$\|\boldsymbol{w}\| = \|\sum_i |w_i|\psi_i(x)\|_{\mathcal{B}}.$$

This norm is finite if and only if the norm of f is finite, i.e., there exist strictly positive constants c and C such that

$$\sqrt{c}\|\boldsymbol{w}\| \leq \|f\|_{\mathcal{B}} \leq \sqrt{C}\|\boldsymbol{w}\|. \tag{4.4}$$

Wavelets are well-known examples of unconditional bases for a variety of spaces, such as Besov spaces.

If we are working in a Hilbert space $\mathcal{H}$, we can replace the induced sequence norm with the ℓ_2-norm (this is a non-trivial step) and state that a basis is unconditional if

$$c\|\boldsymbol{w}\|_2^2 \leq \|f\|_{\mathcal{H}}^2 \leq C\|\boldsymbol{w}\|_2^2, \tag{4.5}$$

4.2 Condition from Finite to Infinite Dimensions

provided that the basis functions are *quasi-normalized* (or *almost normalized*), i.e., there exist strictly positive constants a and A such that

$$a \leq \|\psi_i\|_{\mathcal{H}}^2 \leq A, \forall i \in \mathbb{N}.$$

Such a basis is called a *Riesz basis*. The tightest values of c and C are called *Riesz constants*. Expression (4.5) also reads as: if you manipulate the coefficients (for instance by thresholding them), the effect on the original data is a function which has a norm bounded by the norm of the manipulation in the coefficient domain. You can safely perform some operations in the transformed domain: as long as the operation there is bounded in energy, so will be the effect on the result after reconstruction.

The Riesz representation theorem proves that for each coefficient w_i there exists a function $\tilde{\psi}_i \in \mathcal{H}$ such that

$$w_i = \langle f, \tilde{\psi}_i \rangle.$$

We can write

$$c \sum_i \langle f, \tilde{\psi}_i \rangle \leq \|f\|_{\mathcal{H}}^2 \leq C \sum_i \langle f, \tilde{\psi}_i \rangle.$$

It then holds [57, page 595] that

$$\frac{1}{C} \langle f, \psi_i \rangle \leq \|f\|_{\mathcal{H}}^2 \leq \frac{1}{c} \langle f, \psi_i \rangle. \tag{4.6}$$

This is sometimes taken as an alternative definition of a Riesz basis. The functions $\tilde{\psi}_i$ are the dual basis functions. In wavelet analysis, they correspond to the analysis wavelet basis, whereas the set $\{\psi_i\}$ are the primal basis functions for synthesis of the original signal. Analysis and synthesis are mutually inverse transforms in a biorthogonal wavelet system. The existence of an inverse transform is, however, not enough for a Riesz basis. Every step in a lifting scheme, for instance, is always immediately invertible, but this is no guarantee for stability. The argument based on the Riesz representation theorem can, however, be further refined [19, Remark 1.2]: a (primal) multiresolution basis in a Hilbert space is a Riesz basis if and only if there exists a dual basis, which is biorthogonal to the primal one and which is also a Riesz basis. In other words, non-orthogonal Riesz bases always come in pairs.

Besides the definition of unconditional basis, and Equations (4.5) and (4.6), there is still a fourth way to express what a Riesz basis is. Indeed, it is essentially the next best thing after an orthogonal decomposition. This can be formalized by stating that a Riesz basis is a *topological isomorphism* of an orthonormal basis, i.e., there exists a continuous, bijective transform with continuous inverse that maps a Riesz basis onto an orthogonal basis. Riesz and orthonormal are topologically equivalent.

Bounds of the type (4.6) also exist for stable *overcomplete representations*. Such representations are called *frames* and the associated constants are frame constants. The lower and upper bounds in (4.6) are not enough to guarantee that the system $\{\psi_i\}, i \in \mathbb{N}$ is a basis in the first place; it is an additional condition for stable bases in infinite dimensional vector spaces.

In finite dimensional Hilbert spaces, every basis is a Riesz basis. It is easy to prove that the condition number of a matrix W equals

$$\kappa(W) = \sqrt{\frac{C}{c}},$$

where c and C are the Riesz constants of the basis constituted by the columns of W. The condition number of an invertible discrete wavelet transform matrix for the decomposition of a finite vector of observations is thus always finite. For a lifted wavelet transform, this finite invertibility is guaranteed by the fact that every single lifting step is readily invertible. Yet, it is important to study the asymptotic condition number, i.e., for an increasing number of input observations, as this condition number may become arbitrarily large. This motivates the study of the stability of wavelet expansions in infinite dimensional spaces.

In infinite dimensions, instability in a Schauder basis may occur from the fact that, within a sequence of basis vectors, angles between two vectors can become arbitrarily small, even if the vectors remain independent. In ℓ_2 for instance, one could consider the basis $\{u_i\}, i \in \mathbb{N}$, where $u_1 = e_1$ and

$$u_i = e_1 + e_i/2^i,$$

and $e_i = (0, \ldots, 0, 1, 0, \ldots)$ stands for the ith element of the canonical basis of ℓ_2. This basis is quasi-normalized, yet the angles between two successive vectors and also between the vectors and the first element grow arbitrarily close to zero if $i \to \infty$. The sequence $(1, 1/2, 1/2^2, \ldots) \in \ell_2$ has no representation with square summable coefficients in this basis.

Angles between individual basis vectors in a Hilbert space can be measured by the inner product (dot product) of the normalized vectors. The matrix containing all these inner products is the Gram matrix of the transform. Bounds on the individual elements of the Gram matrix are, however, by no means sufficient to guarantee a stable basis. This can be illustrated in the three-dimensional space, as in Figure 4.3. Although a basis in three dimensions is always a Riesz basis, one of the three basis vectors can be arbitrarily close to the plane spanned by the other two. If there are an infinite number of dimensions then such constructions are easy to make. In other words, one has to check whether any linear combination of basis vectors can come arbitrarily close to any other linear combination of basis vectors.

4.3 Numerical Condition of Wavelet Transforms

Before discussing the stability of a lifted wavelet transform, we first develop some general results on multiscale decompositions.

4.3.1 General Stability Criteria

Since the application of the classical Fourier arsenal on discretized data largely relies on uniform sampling, the stability analysis of unequally spaced data transforms seeks for criteria that are not expressed in terms of Fourier transforms.

4.3 Numerical Condition of Wavelet Transforms

Let $\{\mathcal{V}_j, j \in \mathbb{N}\}$ be a sequence of nested subspaces in an MRA of L_2 and $\Phi_j = \{\varphi_{j,k}, k = 1, \ldots, 2^j\}, j \in \mathbb{N}$ a Schauder basis for $\mathcal{V}_j$. We call the sequence of bases $\{\Phi_j\}$ *uniformly stable* if there exists constants c and C, not depending on j, such that any function $f_j = \sum_{k=1}^{2^j} s_{j,k} \phi_{j,k}$ satisfies

$$c \sum_{k=1}^{2^j} |s_{j,k}|^2 \leq \|f\|^2 \leq C \sum_{k=1}^{2^j} |s_{j,k}|^2.$$

In other words, within the corresponding subspace, the single-scale basis has Riesz bounds that are bounded away from 0 and ∞ uniformly in j. Note that the classical definition of an MRA [57, Definition 7.1] imposes that Φ_0 is a Riesz basis for $\mathcal{V}_0$. The dilation-translation construction of a classical MRA then immediately implies uniform stability for the whole MRA.

A wavelet transform in this MRA can be seen as a change of basis. We call $W^{[J]}$ the transform that synthesizes the wavelet coefficients at the coarsest scales back into fine scale scaling coefficients. More precisely, if $\boldsymbol{w}_j$ is the vector of wavelet coefficients $w_{j,k}$ at scale j, and if $\boldsymbol{s}_J$ is the vector of scaling coefficients $s_{J,k}$ at scale J, then we have

$$\boldsymbol{s}_J = W^{[J]} \begin{bmatrix} \boldsymbol{s}_0 \\ \boldsymbol{w}_0 \\ \boldsymbol{w}_1 \\ \vdots \\ \boldsymbol{w}_{J-1} \end{bmatrix}.$$

We can write

$$W^{[J]} = W_{[J]} \cdot \begin{bmatrix} W^{[J-1]} & 0_{J-1} \\ 0_{J-1} & I_{J-1} \end{bmatrix},$$

where $W_{[j]}$ is the one-level reconstruction transform that maps the wavelet coefficients $[\boldsymbol{s}_{j-1}^T \boldsymbol{w}_{j-1}^T]^T$ onto the scaling coefficients of the next, finer scale and 0_{J-1} and I_{J-1} stand for the square zero matrix and identity matrix of size 2^{J-1} respectively. The following theorem [18, Proposition 2.8] relates the uniform stability of $W^{[J]}$ to the stability of the wavelet basis.

Theorem 4.1 *Suppose that the scaling bases Φ_j are uniformly stable and that the MRA is dense in L_2, i.e.*

$$\overline{\bigcup_{J=0}^{\infty} \mathcal{V}_J} = L_2,$$

and suppose that $W^{[J]}, J = 0, \ldots, \infty$ is a wavelet transform within this MRA, then the corresponding wavelet basis $\{\psi_{j,k}, j \in \mathbb{N}, k = 1, \ldots 2^j\}$ is a Riesz basis (stable basis) if and only if $W^{[J]}$ is uniformly stable, i.e., there exists an upper bound for the sequence of condition numbers $\kappa(W^{[J]})$.

The theorem states that multiscale Riesz stability can be formulated in terms of basis functions, but also in terms of wavelet transforms. Whichever formulation is

88   4. Numerical Condition

used, multiscale Riesz stability is obtained if and only if two aspects are fulfilled [18, 70, Theorem 2.4]:

1. The wavelet *bases* $\{\psi_{j,k}\}$ are uniformly stable within the space $\mathcal{W}_j$ they span.
2. The multiresolution decomposition itself (across scales) is stable. This condition is independent from the choice of basis; it means that there exist constants m and M such that if a function L_2 is decomposed into components at successive scales, i.e., if

$$f = f_0 + \sum_{j=0}^{\infty} g_j \in L_2,$$

with $f_0 \in \mathcal{V}_0$ and $g_j \in \mathcal{W}_j$, then

$$m \left(\|f_0\|^2 + \sum_{j=0}^{\infty} \|g_j\|^2 \right) \leq \|f\|^2 \leq M \left(\|f_0\|^2 + \sum_{j=0}^{\infty} \|g_j\|^2 \right).$$

In the context of finite-element methods, this is an example of what is called *stable subspace splitting* [63, 82].

We remark that these two conditions are both formulated without a reference to uniform stability of the scaling bases. The second condition is generally the harder one. A simple case is the situation where all detail spaces $\mathcal{W}_j$ are orthogonal. The squared norm of f then equals the sum of squared norms of the multiscale components f_j exactly. If in such a situation the wavelets within a scale are not orthogonal, then the construction is called *semi-orthogonal*. Since non-uniform meshes do not allow for basis functions that are simple dilations of each other, orthogonality within a scale is hard to construct.

A necessary, but not sufficient, condition for uniform stability of a wavelet decomposition $W^{[J]}$, is that the one-scale transforms $W_{[j]}$ are uniformly stable [69, Proposition 2.5]. This uniform one-scale stability is again a matter of two aspects [69, Proposition 2.10]:

1. The wavelet *bases* $\{\psi_{j,k}\}$ are uniformly stable within the space $\mathcal{W}_j$ they span.
2. *Uniform complement stability*, i.e., the detail space $\mathcal{W}_j$ is stable complement of $\mathcal{V}_j$ in $\mathcal{V}_{j+1}$, meaning that any function $f_{j+1} \in \mathcal{V}_{j+1}$ can be written as

$$f_{j+1} = f_j + g_j,$$

with $f_j \in \mathcal{V}_j$ and $g_j \in \mathcal{W}_j$, and this decomposition is bounded and boundedly invertible. Moreover, the bounds are independent from j. In other words, the angle between $\mathcal{V}_j$ and $\mathcal{W}_j$ is uniformly bounded away from zero. (The angle between two subspaces is the infimum of angles between any two elements from both subspaces.)

Uniform complement stability is not sufficient for a complete uniformly stable multiresolution decomposition. Intuitively, this can be understood as follows: unless in

4.3 Numerical Condition of Wavelet Transforms

the case of semi-orthogonal wavelets, small deviations from orthogonality may accumulate and result in subspaces that lie arbitrarily close to each other — even if all the individual deviations are uniformly bounded. A typical example is the basis that comes from a lifting scheme without any update step: by subdivision, the corresponding collection of wavelet basis functions at a given scale is simply half of the set of scaling functions at the next, finer scale. In literature on finite-element methods, such a construction is often refered to as a *standard hierarchical basis* [83]. If the prediction step in such a lifted wavelet transform is bounded, then the one-scale transform is, of course, bounded and boundedly invertible. In the long run, however (i.e., if the scale tends to infinity), such a construction is not even a Schauder basis! *An update step is absolutely necessary for stability.* Figure 4.5 illustrates the case of a Haar transform without update. It is easy to see that the zero function (the function which is zero everywhere) can be decomposed in an infinite number of non-trivial ways as a linear combination of these functions. Indeed, we have (assuming here that all functions are normalized to have nonzero values equal to one):

$$\begin{aligned} 0 &= \varphi_{00} - \psi_{00} - \psi_{10} - \psi_{20} - \ldots \\ &= \psi_{00} - \psi_{11} - \psi_{22} - \ldots \\ &= \ldots . \end{aligned}$$

Yet, individual basis functions at any two scales have bounded inner products. Indeed, for L_2 normalized basis functions $\psi_{j,k}$ we have constant inner products of overlapping functions at successive scales, i.e., $\langle \psi_{j,k}, \psi_{j+1,2k} \rangle = 2^{-1/2}$, and vanishing inner products for overlapping functions at non-successive scales, i.e., $\langle \psi_{j,k}, \psi_{00} \rangle = 2^{-j/2} \to 0$. The only remaining class of inner products contains functions with disjunct supports, which are, of course, orthogonal. Moreover, the wavelet basis within each complement space $\mathcal{W}_j$ is orthogonal. Combinations of basis functions at different levels, however, can be arbitrarily close to zero.

4.3.2 Smoothness, Convergence (of Approximation) and Stability

As mentioned before, in classical wavelet theory, stability follows from convergence of the subdivision scheme. Finding the mother (or father) basis functions by subdivision or any other iterative procedure [73] is related to the issue of convergence of general approximations in the given MRA.

Results in approximation theory typically link convergence of approximation to smoothness of the function to be approximated: if that function is smooth in some sense, it can be well approximated and the approximations will also be smooth. The notion of smoothness can be formalized into a smoothness semi-norm [33], or a modulus of smoothness [18]. Results in approximation theory are then typically formulated as two kinds of inequalities:

1. **Jackson or direct estimates** state that the optimal rate of convergence of an approximation f_n of a functiuon f in a Hilbert space is determined by the smoothness semi-norm $|f|_S$ of the objective function f:

4. Numerical Condition

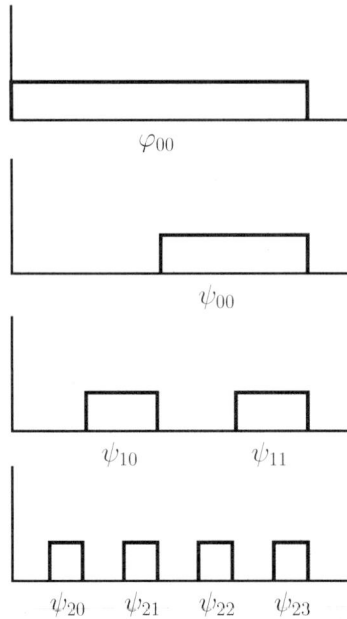

Figure 4.5. The basis functions corresponding to a Haar lifting scheme without update step. These functions do not constitute a Schauder basis, let alone a stable basis.

$$\inf_{f_n \in \mathcal{H}_n} \|f_n - f\|_{\mathcal{H}} \leq Cn^{-r}|f|_{\mathcal{S}}, \tag{4.7}$$

for some positive constants C and r. The infimum is taken over all possible approximations that fall under index n. The space, possibly nonlinear, of these nth-order approximations is denoted as $\mathcal{H}_n$. In our case, this corresponds to the nested scaling spaces $\mathcal{V}_n$. In other words, if the function f is sufficiently smooth, i.e., if its smoothness semi-norm is finite, it can be well approximated.

2. **Bernstein or inverse estimates** state that all possible approximations are smooth as well:

$$|f_n|_{\mathcal{S}} \leq Cn^r \|f_n\|_{\mathcal{H}}, \forall f_n \in \mathcal{H}_n. \tag{4.8}$$

If an approximation in a primal and corresponding dual multiscale decomposition satisfies a Jackson and Bernstein inequality for some appropriate modulus of smoothness, then this biorthogonal system is Riesz stable [19, Theorem 3.3], see also [71, 18, Theorem 3.13]. So, investigation of the approximation behaviour allows — at least in theory — to draw conclusions about stability. We remark that this is a direct statement about the multiscale wavelet representation, it does not involve the wavelet transform from the MRA with scaling functions. Unfortunately, at present, little is known about this approximation behaviour for second-generation wavelets constructed by the lifting scheme. A first step is the convergence analysis of the subdivision scheme on irregular grids and the smoothness of the resulting basis functions. Even on this issue, current results are fragmentary. Cubic polynomial subdivision converges on a wide class of irregular grids, including homogeneous grids (to be defined in Section 4.4) [24, 25]. There is also a connection between convergence of (Lagrange) interpolating subdivision schemes and average interpolating schemes [26], which permits one to extend the convergence results of cubic interpolation towards third-order average interpolation [70]. Convergence of subdivision guarantees uniformly stable scaling bases if the subdivision scheme is bounded [20, Lemma 2.1]. This is a first step towards a stable wavelet decomposition as laid out in Theorem 4.1.

4.4 Numerical Condition of Lifted Wavelet Transforms

We now proceed with the stability analysis of different lifting strategies. From the examples in Section 4.1, it is clear that the irregularity plays a central role in this analysis: algorithms that are stable on regular grids can become problematic on irregularly spaced data. The degree of irregularity can be expressed by a *homogeneity constant*. In one dimension, this homogeneity γ of a multilevel grid is defined as [25, Section 2]:

$$\gamma = \sup_{j,k} \frac{\max(x_{j,k+2} - x_{j,k+1}, x_{j,k} - x_{j,k-1})}{x_{j,k+1} - x_{j,k}}. \tag{4.9}$$

A multilevel grid is called homogeneous if $\gamma < \infty$. In a multiscale triangulation Δ on 2-d data, homogeneity can be defined as $\gamma = 1/\theta_\Delta$, where θ_Δ is the smallest angle in the triangulation [79].

4.4.1 Lifting of Existing Stable Schemes

The effect of a primal (i.e., update) lifting step on the overall stability can be summarized in the following theorem [70, Proposition 3.1]:

Theorem 4.2 *Suppose we have a multiresolution analysis with uniformly stable scaling basis functions and a multiscale decomposition with uniformly stable wavelet basis functions and uniformly stable complements. If the update operations on all scales j are uniformly bounded in norm, then the lifted wavelet bases again are uniformly stable and the new multiscale decomposition has again uniformly stable complements.*

Similar results hold for dual lifting (i.e., prediction) steps. In other words, lifting with bounded operations preserves uniform single scale basis stability and complement stability. Unfortunately, this is not sufficient for overall stability of the multiscale decomposition, as discussed in Section 4.3.1. The theorem is, however, important in the sense that uniformly bounded operations are not a built-in feature of the lifting scheme. For instance, it is easy to construct irregular grids on which interpolation prediction coefficients can become arbitrarily large, as we discuss below. Taking into account the conditions of this theorem may result in transforms that do have condition numbers that are sufficiently small for practical use.

4.4.2 Numerical Condition and Primal Vanishing Moments

Dual vanishing moments play a crucial role in the quality of approximation of a wavelet expansion, which is important in applications such as compression, but also in threshold methods for smoothing. If the wavelet analysis has a high number of vanishing moments, smooth sections — i.e., sections that can be well approximated by polynomials — generate small, neglectible coefficients. Primal vanishing moments, on the other hand, play a more direct role in stability. An intuitive argument for primal vanishing moments goes as follows. If we assume that a coarse scale wavelet basis function is smooth, it can be (locally) well approximated by a polynomial. The inner product of this coarse scale wavelet function, with some finer scale function at the same location, is then small. This relationship between primal moments and stability is, however, far from exclusive. The inner products of different basis functions (i.e., the entries of the Gram matrix of the basis) are not sufficient to characterize the stability in the first place.

Nevertheless, in many discussions, one primal vanishing moment shows up as an interesting, necessary, or sufficient condition in stability analyses. In classical continuous wavelet analysis, and in frame theory, the *admissibility* of the mother wavelet is a necessary condition for a numerically stable reconstruction [22, Sections 1.3, 2.4, Theorem 3.3.1]. This admissibility basically reduces to the requirement of a zero integral, i.e., a first vanishing moment. It is also an interesting starting point to establish a Jackson estimate [18, Section 3].

In a discrete MRA on an equidistant grid, the first vanishing moment follows from the fact that a constant function can be represented exactly in a scaling basis at arbitrary scale. If this holds for the dual scaling basis as well, then the primal wavelet function has to have a vanishing integral, as primal wavelets are connected to dual scaling functions [73].

This scaling basis property is known as the *partition of unity*. It shows up for at least two reasons. First, it is a necessary condition in the convergence of any numerical, iterative solution of the two-scale (or dilation) equation. We could, for instance, solve the two-scale equation by running it backwards (i.e., from coarse to infinitely fine resolution) starting from a Kronceker sequence. This is subdivision. If we run subdivision on a regular grid, the iterative procedure is characterized by a single matrix, whose eigenvalues have to be smaller than one. The stability analysis of that matrix leads to conditions that imply partition of unity.

Second, partition of unity is a necessary condition for a Riemann integrable scaling function to even *satisfy* the two-scale (or dilation) equation. Indeed, if $\varphi(x)$ is Riemann integrable and satisfies

$$\varphi(x) = \sum_{k \in \mathbb{Z}} c_k \varphi(2x - k),$$

then

$$\sum_{k \in \mathbb{Z}} \varphi(x - k) = 1.$$

This can be proven as follows: consider

4.4 Numerical Condition of Lifted Wavelet Transforms

$$A_J(x) := \sum_{k \in \mathbb{Z}} \varphi\left(\frac{x-k}{2^J}\right).$$

So we have to prove that $A_0(x) = 1$. First note that $A_0(x)$ is periodic: $A_0(x+l) = A_0(x)$. We only have to consider $0 \leq x \leq 1$. Now we have:

$$\begin{aligned}
A_J(x) &= \sum_{k \in \mathbb{Z}} \varphi\left(\frac{x-k}{2^J}\right) \\
&= \sum_{k \in \mathbb{Z}} \sum_{l \in \mathbb{Z}} c_l \varphi\left(\frac{x-k}{2^{J-1}} - l\right) \\
& \sum_{l \in \mathbb{Z}} c_l \sum_{k \in \mathbb{Z}} \varphi\left(\frac{x-k-2^{J-1}l}{2^{J-1}}\right) \\
& \text{suppose } k + 2^{J-1}l = n \\
&= \left(\sum_{l \in \mathbb{Z}} c_l\right) \sum_{n \in \mathbb{Z}} \varphi\left(\frac{x-n}{2^{J-1}}\right) \\
&= 2 A_{J-1}(x) = 2^J A_0(x).
\end{aligned}$$

So

$$\begin{aligned}
A_0(x) &= 2^{-J} A_J(x) = \lim_{J \to \infty} 2^{-J} A_J(x) \\
&= \lim_{J \to \infty} \sum_{l \in \mathbb{Z}} 2^{-J} \varphi\left(\frac{l}{2^J} + \frac{x}{2^J}\right).
\end{aligned}$$

Since $0 \leq x \leq 1$, we have

$$\frac{l}{2^J} + \frac{x}{2^J} \in \left[\frac{l}{2^J}, \frac{l+1}{2^J}\right]$$

So, we have the limit of a Riemann-sum, which is the integral:

$$A_0(x) = \int_{-\infty}^{\infty} \varphi(s) ds = 1.$$

This proof, however, depends heavily on the fact that translations of dilations of a single father function constitute the basis at successive scales. It cannot be extended towards non-equispaced grids. The necessity of a primal vanishing moment in general has to do with something else.

In Section 4.3.1, Figure 4.5, we argued that a lifting scheme without update step leads to wavelets that do not constitute a Schauder basis. This can be extended to wavelets without a vanishing integral, and the Haar case is still illustrative. Suppose indeed that we have a nontrivial function $f_J \in V_J$ and use this as an approximation of the zero function. For ease of reasoning, if we are working with functions on

94   4. Numerical Condition

an unbounded domain, we take an approximation with compact support. We now construct a function $f_{J+1} = f_J - g_J \in \mathcal{V}_{J+1}$ such that f_{J+1} is the best possible approximation of the zero function in L_2. In other words, $g_J = \sum_{k \in \mathbb{Z}} a_{J,k} \psi_{J,k}$ is the least-squares approximation of f_J in the complement $\mathcal{W}_J$. If the approximation f_J converges to zero, we have constructed a nontrivial decomposition of the zero function, and so the system $\{\varphi_{0,k}, \psi_{j,k}, j \in \mathbb{N}, k \in \mathbb{Z}\}$ cannot be a Schauder basis. In the case of Haar scaling functions, suppose that we do add a one-tap update, but this update is not designed to create a vanishing moment, we end up with basis functions that are block wavelets without zero integral, i.e., without loss of generality, $\psi_{0,0}(x) = -b$ on $[0, x_1]$ and $\psi_{0,0}(x) = 1$ on $[x_1, 1]$. Let $\Delta_1 = x_1$ and $\Delta_2 = 1 - x_1$. Then, take $f_0(x) = \varphi_{0,0}(x) = \chi_{[0,1]}(x)$ and $f_1(x) = \varphi_{0,0}(x) - a_1 \psi_{0,0}(x)$. For $a_1 = (\Delta_2 - b\Delta_1)/(\Delta_2 b^2 + \Delta_1)$, we find $\|f_1\|_2^2 = 1 - (\Delta_2 - b\Delta_1)^2/(\Delta_2 b^2 + \Delta_1)$, which is strictly smaller than 1 except when $b = 1$. The procedure can then be repeated on the subintervals $[0, x_1]$ and $[x_1, 1]$, leading to an L_2-convergent sequence $f_j \to 0$. Note that, in this example, the individual wavelets even constitute an orthogonal set: nonzero integrals do not make small angles as such.

In a general multiscale decomposition, we find for the least-squares coefficients $a_{J,k}$ (i.e., by plugging in the normal equations into the expression $\|f_{J+1}\|_2^2 = \|f_J\|_2^2 - 2\langle f_J, g_J \rangle + \|g_J\|_2^2$) that

$$\|f_{J+1}\|_2^2 = \|f_J\|_2^2 - \sum_{k \in \mathbb{Z}} a_{J,k} \langle f_J, \psi_{J,k} \rangle.$$

If the $\psi_{J,k}$ constitute an orthogonal system within the complement space $\mathcal{W}_{J,k}$ at scale J, it holds that

$$\|f_{J+1}\|_2^2 = \|f_J\|_2^2 - \sum_{k \in \mathbb{Z}} \langle f_J, \psi_{J,k} \rangle^2 / \|\psi_{J,k}\|_2^2.$$

Otherwise, the system can be orthogonalized within each scale, leading to similar conclusions: there are $\mathcal{O}(2^J)$ nonzero coefficients, each having a magnitude of $\mathcal{O}(2^{-J} \|f_J\|_2^2)$. If the first moment (the integral of $\psi_{J,k}$) were to be zero, the terms would vanish more quickly. As a result for the case of nonzero integrals, $\|f_{J+1}\|_2^2$ is a fraction, strictly smaller than one, of $\|f_J\|_2^2$, and so we have constructed a converging, nontrivial decomposition of the zero function.

This first primal moment is reported to be crucial in practical applications [70, Section 5.3].

4.4.3 Stabilizing Updates

As discussed in Section 2.2.3, Figure 2.6, the update step creates the wavelet *basis functions*, whereas the prediction takes care of the wavelet (detail) *coefficients*. Since stability is a matter of basis functions in the first place, the primal lifting step is the first candidate to take care of stability. The standard update step, on regular grids, is designed to maximize the number of primal vanishing moments. As discussed in Section 4.4.2, one primal vanishing moment is a prerequisite for a unique decomposition, so all subsequent criteria in primal filter design are subject to the side condition of at least one vanishing moment.

Suppose $A_{j,k,l}$ are the update coefficients at scale j, location k. The new, lifted basis function is then

$$\psi_{j,k} = \psi_{j,k}^{[0]} + \sum_i A_{j,k,l} \varphi_{j,l}.$$

If one or more of these coefficients are large, i.e., if $A_{j,k,l} \gg \|\psi_{j,k}^{[0]}\|/\|\varphi_{j,l}\|$, then the lifted wavelet $\psi_{j,k}$ nearly falls within the vector space spanned by its neighbouring scaling functions at the same scale. This creates a detail space which is far from orthogonal to the coarse scaling space. Large update coefficients result in a considerable overlap of scaling and wavelet function at a given scale. When the scaling functions are further decomposed into a wavelet basis at coarser scales, the immediate correlation between basis functions becomes hidden. The bad condition is then hard to detect in advance and hard to localize. The standard update spends all degrees of freedom on vanishing moments. On irregular grids, and following an interpolating prediction, this may result in unboundedly large update coefficients [78, Section 4.2]. Relaxing on the number of vanishing moments (or extending the update lattice) creates some degrees of freedom to minimize the magnitude of the update coefficients [50].

Instead of this purely algebraic approach on the coefficients only, one may consider a semi-orthogonalization of the spaces $\mathcal{V}_j$ and its complement $\mathcal{W}_j$ in $\mathcal{V}_{j+1}$. It suffices to orthogonalize the basis functions $\psi_{j,k}$ with respect to the scaling functions. Since all polynomials with degree smaller than the number of dual vanishing moments are in $\mathcal{V}_j$, semi-orthogonalization would automatically yield a number of primal vanishing moments equal to the dual number. Unfortunately, full semi-orthogonalization requires an infinite number of update coefficients (degrees of freedom) and, hence, the wavelets do not possibly have compact support.

One solution for this is to chose for each wavelet function $\psi_{j,k}$ a limited number of scaling functions $\varphi_{j,l}$. The update then creates a wavelet orthogonal to the space spanned by these selected scaling functions. Since scaling functions far away from a given wavelet have little overlap with this wavelet, this local semi-orthogonalization performs quite well in resolving instabilities [70]. Finally, an additional update corrects for the first vanishing moment.

An alternative for this two-stage solution (local semi-orthogonalization followed by vanishing integral update), is to solve an overdetermined system for both local semi-orthogonalization and vanishing moment. The vanishing moment condition is strict, so, after elimination of this equation, the remaining conditions are solved in a least-squares sense [79].

4.4.4 Prediction

Although the dual lifting step does not operate on the wavelet basis functions directly, it is important to keep an eye on stability when adding a prediction step to a lifting scheme. Indeed, predictions change the two-scale (refinement) equation and, hence, have an impact on the scaling basis functions, and so on the multiresolution subspaces $\mathcal{V}_j$. The wavelet basis functions have to be constructed within these subspaces. Uncareful prediction may introduce severe instabilities, which may be reinforced by a subsequent primal lifting step [78, Section 4.2].

It is interesting to have prediction coefficients between zero and one. Indeed, if the prediction has at least one (dual) vanishing moment, then we know that the sum of the prediction coefficients $a_{j,k,l}$ equals 1. Given that all coefficients are positive, we can write

$$\|\boldsymbol{a}_{j,k}\|_1 = \sum_l |a_{j,k,l}| = 1.$$

This prevents errors ε on the input (i.e., even samples) from blowing up, since the prediction error η satisfies

$$|\eta_k| \leq \sum_l |a_{j,k,l}||\varepsilon_l| \leq \sum_l |a_{j,k,l}| \max_l |\varepsilon_l| = \|\boldsymbol{a}_{j,k}\|_1 \cdot \|\varepsilon\|_\infty.$$

If the prediction is bounded, then conditions on the update could be relaxed. Bounded prediction coefficients are out of the question in the case of, for instance, interpolating prediction on arbirtrary grids, except for the linear prediction case. Indeed, consider the grid in Figure 4.6, where even grid points are marked with an asterisk and odd points with a circle. Some of the points are close to each other so that the symbols overlap. The grid was generated as samples from a uniform distribution. In Figure 4.6(a) we see the cubic prediction at work: prediction coefficients in a given odd point x_{odd} are found by cubic interpolation of a Kronecker sequence with a one in a neighbouring even point. The figure shows two out of the four coefficient constructions for this odd point. Since the neighbouring evens on the left are much closer to each other than to the odd point where the cubic interpolant is evaluated, the coefficients may reach far outside the safe $[0, 1]$ interval. After one step of the subdivision scheme, the result is scaling functions with heavy side lobes, as shown in Figure 4.6(b). When using this scaling function in an update step, the resulting, lifted wavelet function inherits this side lobe, and so it has a major feature in common with the scaling function, leading to an unwanted overlap. This effect is even enhanced by the fact that such heavy tailed basis functions generally require large update coefficients to annihilate each other's effect. This problem arises from the fact that small-scale phenomena are somehow extrapolated towards larger scales. This scale mixing is the major source of instabilities in lifted wavelet transforms. Note that it cannot happen in the lifting implementation of the CDF 2,2 basis, i.e., when using linear prediction followed by a two-taps update. The prediction coefficients are always in the $[0, 1]$ interval. In other words, the two even neighbours are always closer to the odd point where the prediction is evaluated than to each other, so scale mixing is impossible. The CDF 2,2 transform is always stable [47]. In two dimensions, the situation becomes more dramatic: even if a prediction scheme only takes immediate neighbours into account, these neighbours can be closer to each other than to the point where the prediction is evaluated; see Figure 4.7.

In order to suppress large prediction coefficients, interpolation or average interpolation could be replaced by more robust prediction schemes. Least-squares pre-

4.4 Numerical Condition of Lifted Wavelet Transforms   97

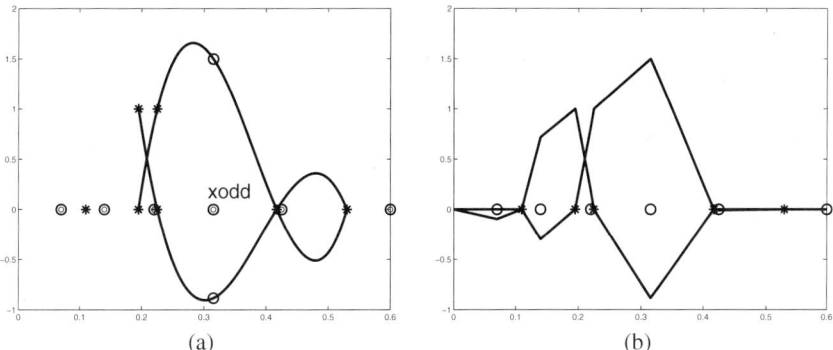

Figure 4.6. Mixing of scales causes large prediction coefficients (a), and, by subdivision, it creates scaling functions with heavy side lobes (b).

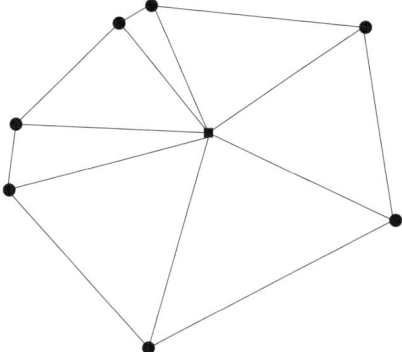

Figure 4.7. In two dimensions, scale mixing occurs even when a prediction scheme only uses immediate neighbours. This is impossible in one dimension: linear prediction is always stable.

diction could be interesting, but has the drawback that, in general, the predicted value in an odd point does not converge to the value of an even point, even if the odd point approaches the even point as in Figure 4.8. A weighted least-squares prediction with weights proportional to the inverse distance could be a solution.

In two dimensions, the linear prediction of the CDF 2,2 can be extended to so-called natural interpolation [68]. As illustrated in Figure 4.9, this natural extension uses the unsigned areas of the Voronoi cells of the left out vertex, indexed by 0, and its neighbours before and after the removal. By removing the vertex, its cell is distributed in parts among its neighbours. The prediction coefficients are simply the proportions of the central cell's area, taken by the neighbours:

$$\beta_{j,k} = \frac{A_{j+1,0,k}}{A_{j+1,0}},$$

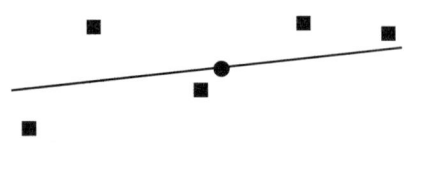

Figure 4.8. Linear least-squares prediction of an odd point (circle) by four neighbouring even points (squares —other odd points not indicated on the figure). Because the corresponding subdivision scheme for the scaling functions is not interpolating and at the same time does not update existing, even values, this unweighted least-squares prediction leads to a discontinuity if the odd point comes arbitrarily close to one of the evens.

where $A_{j+1,0} = \sum_{l \in \partial_{j+1} 0} A_{j+1,0,l}$ is the unsigned area of the central cell (i.e. the Voronoi cell of vertex 0 left out at scale j) and $A_{j+1,0,k}$ is the area of the part of this central cell assigned to neighbour k. The prediction coefficients sum up to one, so the scheme clearly has at least one vanishing moment: a constant function is predicted perfectly. There is also a second vanishing moment, the proof of which is non-trivial [67]. The scheme is interpolating: if the central vertex 0 coincides with one of the neighbours, its entire cell will be annexed to that neighbour's cell. It is an example of piecewise rational interpolation. Indeed, the cell areas are polynomial functions of the coordinates of the cell boundary points. Those points in their turn are rational functions of the sample vertex coordinates. The coefficients in the rational expression of the interpolant change whenever a change in sample locations causes a change in the topology of the triangulation. The resulting piecewise rational function is, however, continuous in all points. In one dimension, this piecewise rational function reduces to a simple linear polynomial.

The prediction coefficients are positive and smaller than one, which ensures a stable transform, at least on finite grids.

Boundary Treatment. In order to maintain the same number of dual vanishing moments near the boundary, the prediction method is forced to use an unbalanced set of even points, as illustrated in Figure 4.10.

Since there are not enough even points between the boundary and an odd point close to the boundary, the prediction in this odd point is based on even points of which less than half is situated between the odd point and boundary and more than half is situated on the other side of the odd point. As a consequence, the same predictor with exactly the same even points is being evaluated for the prediction of several odd points. This has a negative impact on the numerical condition of the transform, since small errors in those even points proliferate in more than the usual number of detail coefficients.

A more serious side effect of keeping the same number of vanishing moments near the boundary is the appearance of basis functions with heavy tails towards the boundary. Indeed, the first odd points in a subdivision stage use even neighbours that are further away than usual. If one of these evens is a nonzero value of a basis

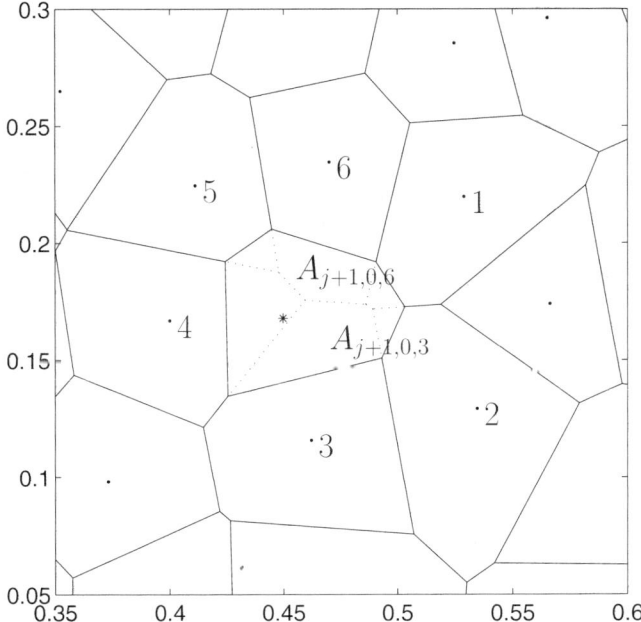

Figure 4.9. Sibson's natural interpolation. The function value $s_{j+1,0}$ in vertex 0 is predicted by a linear combination of its neighbouring values: $\sum_{k \in \partial_{j+1} 0} \beta_{j,k} s_{j+1,k}$, where $\beta_{j,k}$ is the proportion of the cell's area, assigned to neighbour k after removing vertex 0 and its cell:
$$\beta_{j,k} = \frac{A_{j+1,0,k}}{\sum_{l \in \partial_{j+1} 0} A_{j+1,0,l}}.$$

function, then that value causes a nonzero subdivision in those odd points near the boundary. The result is illustrated in Figure 4.11.

4.4.5 Splitting Strategies

Existing results on stability of second-generation wavelets [70, 79] are formulated in terms of the homogeneity of the grid. An inhomogeneous grid can, therefore, still lead to arbitrarily unstable decompositions. As illustrated in Figure 4.6, *scale mixing* is the major source of instabilities. In order to prevent the lifting filters from mixing scales, it is interesting to redesign the splitting stage. Instead of a fixed splitting into evens and odds, the splitting stage can also be made grid dependent.

A first methodology in reconsidering the splitting stage is to start from the classical even-odd subdivision, followed by a careful examination of all the odd indexed locations. If such an odd point lies in a position where it cannot be predicted in a stable way by its even neighbours, the point is added to the set of evens and its prediction is postponed until the scheme reaches a coarser scale where this point

100   4. Numerical Condition

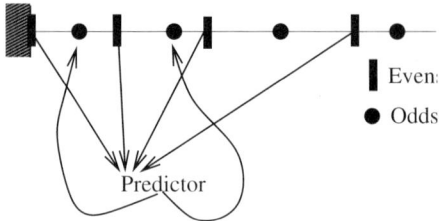

Figure 4.10. Cubic prediction near the boundary. In order to maintain the same number of dual vanishing moments near the boundary, the prediction method is forced to use an unbalanced set of even points. As a consequence, the first odd point is predicted by exactly the same set of evens as the second odd point. This situation is unfavourable for numerical condition.

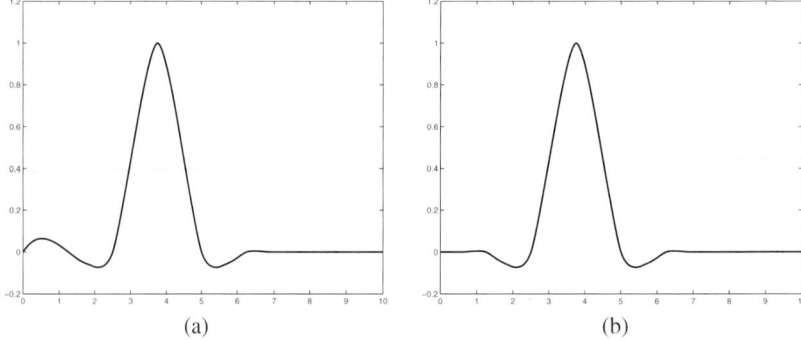

Figure 4.11. Boundary effects for cubic interpolating subdivision, illustrated here on a regular grid. (a) If the algorithm keeps the same number of vanishing moments (namely three) near the boundary, prediction near the boundary has to involve an unbalanced set of points towards the interior. As a consequence, a nonzero value in a step of a subdivisuion scheme is 'observed' by (odd) points near that boundary that otherwise would not look so far. The prediction values are nonzero as well, leading to heavy tails near the boundary. The effect may be even more dramatic on irregular grids. (b) If the prediction remains balanced near the boundary, the number of vanishing moments near the boundary is lower, but a basis function in the interior is not 'observed' by the boundary.

can be predicted without mixing fine and coarse scales. Moving points from the odd set to the even sets may, in its turn, make the location of other odd points less favourable. Also, we have to make sure that no grid becomes 'unsplittable'. All these side conditions result in a quite complicated algorithm [78]. Also, in general 2-d applications, it is impossible to start from a regular even-odd-like subdivision, since such a subdivision does not exist.

A second approach is based on the observation that the construction of a dyadic MRA on irregular data is in fact impossible. In order to deal with all possible locations, scale should be a continuous notion. This leads to an algorithm where in every step only one point is predicted and all other values are considered as 'evens'. Concentrating on one coefficient in each step, it is easier to keep scale mixing un-

4.4 Numerical Condition of Lifted Wavelet Transforms

der control, especially in two dimensions, where scale mixing is more prominent and starting from regular subdivision is generally (except for tensor-product grids) impossible. Nevertheless, a careful elaboration is also necessary for this routine. A straightforward implementation could be to select, in every step, the point that lives on the smallest scale, i.e., the point that is nearest to those neighbours that will serve in the prediction of the point's observed value. Figure 4.12 shows what may happen with this scheme. The initial grid is regular; so, in the first, second and following steps, there are many candidates to be predicted. If we select the leftmost of those candidates first, we run into trouble after the second step, since the selection procedure itself creates a grid with small intervals next to large gaps.

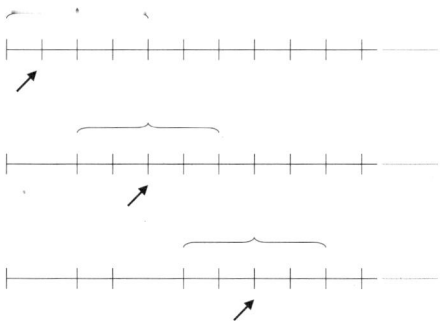

Figure 4.12. Uncareful selecting of one point in each scale may create additional irregularity.

A third approach proceeds in two steps: first the order of point selections is established from coarse to fine scales (top down) and after that, the actual transform (predictions and updates) is carried out according to the prescribed order. In one dimension, this *regridding* or *remeshing* procedure starts from two points on the boundary of the interval. Next, it finds the observation closest to the middle point of the interval between those two end points. This procedure is repeated until all points have been selected.

5. Applications of Nonlinear Lifting in Imaging

In this last chapter we discuss some recently developed applications of nonlinear lifting in various fields of image processing. Since imaging requires approaches in two dimensions, the reader will also meet aspects of lifting that were not dealt with before in the previous chapters.

5.1 Image Retrieval Using Adaptive Lifting

Content-based image retrieval is the process of retrieving desired images from a large collection of images on the basis of features, e.g. similar shape and colour. The extraction process should be automatic, so without any human interference. This is a requirement that is difficult to realize, but which is of great value for applications like image-based browsing on the Internet or in electronic catalogues. In this section we consider such an automatic retrieval algorithm using a database of greyscale images of objects against a background of texture. Then, given an image of an object appearing in the database, the problem is to identify all other images in the database containing the same object irrespective of translation, rotation or re-sizing of the object, lighting conditions and the background texture.

A classical approach to the problem of recognizing similar images is by using moment invariants. Hu introduced in his classical paper [45] invariant quantities based on calculating moments of an image and combining them into polynomials. These polynomials can be used as statistical quantities to identify similar images, i.e., obtained by transforming one single original image (mostly affine transforms). However, Hu's approach only works for images of crisp objects with a neutral background.

To deal with the problem of different backgrounds one may use a filtering process as preprocessing step. In Do et al. [34] the classical wavelet transform was used to preprocess the images. Here, we propose to use adaptive lifting as discussed in Chapter 3. The retrieval process will be based on computing invariant polynomials of the detail coefficients after various successive lifting steps.

5.1.1 Quincunx Lifting

We introduce the Quincunx lifting scheme for decomposing images represented by an (infinite) matrix $I_{k,m}$. Here, each entry $I_{k,m}$ denotes the value of a pixel at

104   5. Applications of Nonlinear Lifting in Imaging

"position"(k, m) in the matrix. Instead of splitting all data into even and odd entries as we did in the 1-d lifting scheme, we split all pixels into red-labelled entries, in the case that k mod $2 = m$ mod 2, and in black-labelled entries otherwise, as depicted in the left-hand side of Figure5.1. This division is also called "checkerboard"or "red black"division.

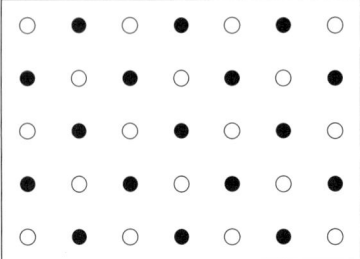 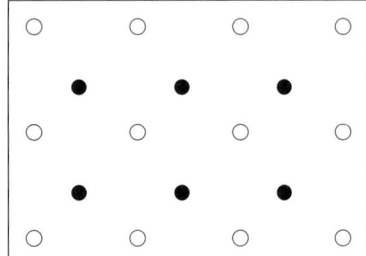

Figure 5.1. The red(○) black(●) division of an image in the first lifting step (left) and the second step (right).

In image processing, the lattices consisting of all red or black spots is well known as the quincunx lattice. Within the lifting scheme the pixels on the red spots (○) are used to predict the samples on the black spots (●), which means that the red-labelled entries act as $s_{j+1,2k+1}$ and the black ones as $s_{j+1,2k}$ as compared with what has been described in the previous chapters. After updating we obtain a detail image on the black spots and an updated approximation image on the red spots. This approximation image can again act as input for the lifting scheme to obtain a detailed image at a coarser level of resolution. For splitting the remaining red spots we rotate the approximation image over 45 degrees and repeat the red black division once more yielding the splitting as shown in the right-hand side of Figure 5.1. Filtering based on this repeated subdivision is known as the red black transform by Uytterhoeven and Bultheel [77]. For convenience we will only describe one lifting step, consisting of prediction and updating of the image pixels. Therefore, we will omit the index j indicating the resolution level and only use I, I_r and I_b, denoting the image and its division on the red and black spots respectively.

As filters in the scheme we take polynomial interpolation filters in two dimensions. Moreover, for primal and dual lifting we take so-called Neville filters of order N and $\widetilde{N}$ respectively see [72]. Kovačević and Sweldens [54] showed that lifting schemes with N primal and $\widetilde{N}$ dual vanishing moments ($N \leq \widetilde{N}$) can always be built using Neville filters of order $\widetilde{N}$ for prediction and half the adjoint of an order N Neville filter as update. As an example we mention the second-order prediction and update filters acting on the original image I

$$(PI)_{k,m} = [I_{k-1,m} + I_{k,m-1} + I_{k+1,m} + I_{k,m+1}]/4, \ k \bmod 2 \neq m \bmod 2,$$
$$(UI)_{k,m} = [I_{k-1,m} + I_{k,m-1} + I_{k+1,m} + I_{k,m+1}]/8, \ k \bmod 2 = m \bmod 2.$$

Obviously, in the prediction step we predict the value of a black-labelled pixel by averaging the values of the four nearest red-labelled neighbouring pixels. For the update we do the same by interchanging red and black spots and dividing the result by 2.

The Neville filters we intend to use for prediction and update can be written as

$$(P_{\widetilde{N}} I)_{k,m} = \sum_{(n,l) \in S_{\widetilde{N}}} a^{\widetilde{N}}_{(n,l)} I_{k+n,m+l}, \; k \bmod 2 \neq m \bmod 2, \quad (5.1)$$

with $S_{\widetilde{N}}$ a subset of $\{(n,l) \in \mathbb{Z}^2 \mid (n+l) \bmod 2 = 1\}$ and $a^{\widetilde{N}}_{n,l}$, $(n,l) \in S_{\widetilde{N}}$, a set of coefficients in $\mathbb{R}$, depending on $\widetilde{N}$. We define $P_{\widetilde{N}}$ on I instead of I_r. However, the contstraints on which entries the prediction filter is acting on clearly show the difference between the red- and black-labelled pixels. The reason for defining $P_{\widetilde{N}}$ on I will be clarified in Section 5.1.3. For U we have

$$(U_N I)_{k,m} = 1/2 \sum_{(n,l) \in S_N} a^N_{n,l} I_{k+n,m+l}, \; n \bmod 2 = l \bmod 2, \quad (5.2)$$

with S_N depending on the number of required primal vanishing moments N. For several elements in S_N the coefficients $a^N_{n,l}(s)$ attain the same values. Therefore, we take these elements together in subsets of S_N, i.e.,

$$\begin{aligned}
V_1 &= \{(\pm 1, 0), (0, \pm 1)\}, \; V_2 = \{(\pm 1, \pm 2), (\pm 2, \pm 1)\}, \\
V_3 &= \{(\pm 3, 0), (0, \pm 3)\}, \; V_4 = \{(\pm 2, \pm 3), (\pm 3, \pm 2)\}, \\
V_5 &= \{(\pm 1, \pm 4), (\pm 4, \pm 1)\}, \; V_6 = \{(\pm 5, 0), (0, \pm 5)\}, \\
V_7 &= \{(\pm 3, \pm 4), (\pm 4, \pm 3)\}.
\end{aligned} \quad (5.3)$$

Table 5.1 indicates the values of all $a^N_{n,l}$, $(n,l) \in V_k$, for different values of N (two to eight) when using quincunx Neville filters see [54].

Table 5.1. Quincunx Neville filter coefficients

order N	V_1	V_2	V_3	V_4	V_5	V_6	V_7
2	1/4	0	0	0	0	0	0
4	10/32	$-1/32$	0	0	0	0	0
6	$87/2^8$	$-27/2^9$	2^{-8}	$3/2^9$	0	0	0
8	$5825/2^{14}$	$-2235/2^{15}$	$625/2^{16}$	$425/2^{15}$	$-75/2^{16}$	$9/2^{16}$	$-5/2^{12}$

We observe that $S_8 = V_1 + \cdots + V_7$ and so a 44 taps filter is used as prediction/update if the required filter order is 8. For an illustration of the Neville filter of order 4 see Figure 5.2. Here, the numbers 1, 2 correspond to the values of the filter coefficients as given in V_1 and V_2 respectively at that position. The left-hand filter

106   5. Applications of Nonlinear Lifting in Imaging

can be used when using the lifting scheme on the image I, while the right-hand filter corresponds to the entries that are left in I_r, which is used for lifting the second time.

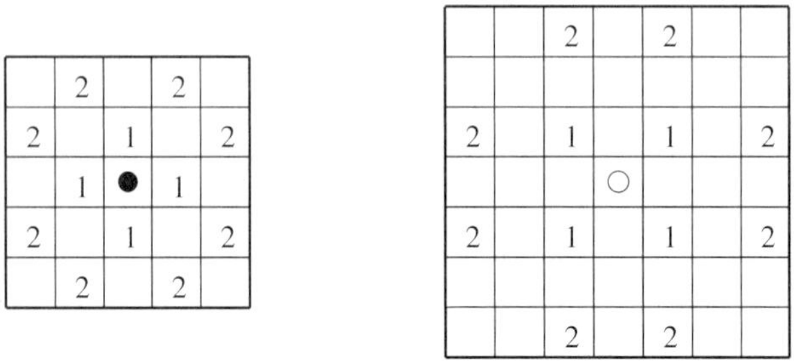

Figure 5.2. Neville filter of order 4 for first time lifting (left) and repeated lifting (right). The numbers 1 and 2 correspond to the positions of the filter taps of V_1 and V_2.

5.1.2 Adaptive Lifting

When using the lifting scheme or a classical wavelet approach, the prediction/update filters or wavelet/ scaling functions are chosen in a fixed fashion. Generally, they can be chosen in such way that a signal is approximated with very high accuracy using only a limited number of coefficients. As noted in Chapter 3, discontinuities mostly give rise to large detail coefficients, which is undesirable for applications like image compression. In this application, large detail coefficients near edges in images are highly desirable, since they can be identified with the shapes of objects we want to identify. However, they are undesirable if such large coefficients are related to the background of the image. This situation occurs if a small filter is used on a background of texture that contains irregularities locally. In this case a large smoothing filter gives rise to small coefficients for the background.

These considerations lead to the idea of using space-adaptive lifting as described for the 1-d case in Figure 3.5 in Chapter 3. Translated to images, this means that high-order interpolation filters will be used for prediction of the background, while prediction of the edges is established with a low-order prediction filter. The filters we use to realize this prediction are Neville filters of order $\widetilde{N} \in \{2, 4, 6, 8\}$ on the quincunx grid. The decision map for determining the prediction filter can be based on both the data on the red spots (I_r) and the black spots (I_b), since only detail coefficients are needed and we do not have to bother about perfect reconstruction. Adaptivity in the update step is not necessary. Therefore, the update filter is taken fixed similar to the lowest order prediction filter, i.e., half the Neville filter of order 2, yielding $N \le \widetilde{N}$ for all possible $\widetilde{N}$. Resuming these considerations, we end up with a lifting scheme as shown in Figure 5.3.

The decision map will be based on the relative local variance as given in (3.11), but extended to two dimensions and based on the whole images $I = I_r + I_b$, yielding

5.1 Image Retrieval Using Adaptive Lifting

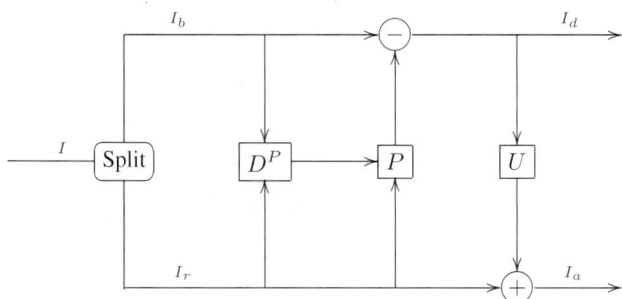

Figure 5.3. Generating coefficients via adaptive lifting: split an image into red- (I_r) and black-labelled pixels (I_b), predict and update to generate a detailed image (I_d) and an approximation image (I_a).

$$\mathrm{RLV}[I]_{k,m} = \sum_{l=k-L}^{k+L} \sum_{n=m-L}^{m+L} (I_{l,n} - \overline{\mu_{k,m}})^2 / \mathrm{var}\,(I), \qquad (5.4)$$

with

$$\overline{\mu_{k,m}} = \sum_{l=k-L}^{k+L} \sum_{n=m-L}^{m+L} I_{l,n}/(2L+1)^2. \qquad (5.5)$$

We take $L = 5$, since with this choice all $I_{l,n}$ that are used for the prediction of $I_{k,m}$ contribute to the RLV for (k,m), even for the eighth-order Neville filter. Before analyzing an image I through the scheme, we first compute the RLV for each pixel (k,m). Next the value of the decision map D^P is calculated for each $I_{k,m}$. This decision map is given by

$$D^P[I_r, I_b]_{k,m} = \begin{cases} 2, & \text{if } \Delta_1 \leq \mathrm{RLV}[I]_{k,m}, \\ 4, & \text{if } \Delta_2 \leq \mathrm{RLV}[I]_{k,m} < \Delta_1, \\ 6, & \text{if } \Delta_3 \leq \mathrm{RLV}[I]_{k,m} < \Delta_2, \\ 8, & \text{if } \mathrm{RLV}[I]_{k,m} < \Delta_3, \end{cases} \qquad (5.6)$$

with $I = I_r + I_b$ and Δ_1, Δ_2 and Δ_3 given positive threshold values. For these threshold values we take multiples of the mean of all computed RLV ($\mu(\mathrm{RLV})$). Test results with databases similar to the one we are dealing with here, have shown that $\Delta_1 = 2\,\mu(\mathrm{RLV})$, $\Delta_2 = 1.5\,\mu(\mathrm{RLV})$ and $\Delta_3 = \mu(\mathrm{RLV})$ yield good results in our application. The prediction filter to be used for $I_{k,m}$ is then given by P_{D^P} as in (5.1). In Figure 5.4 we have depicted an image (left) and its decision map D^P based on the RLV (right). The colour in the right-hand figure indicates the value of the map from black ($D^P = 2$) to white ($D^P = 8$), which covers most regions of the image.

108   5. Applications of Nonlinear Lifting in Imaging

a) original image   b) decision map (RLV)

Figure 5.4. (a) An object on a wooden background and (b) its decision map: white=eighth-order prediction filter, black=second-order prediction filter.

Although at this stage only half the entries of I are predetermined to be used for prediction, the next section will show different. It will turn out that for several desired invariant measures it will be necessary to predict and update at all entries.

5.1.3 Redundant Lifting

Although both traditional wavelet analysis and the lifting scheme yield detail and approximation coefficients that are localized in scale and space, they are not both translation invariant. Therefore, lifting translated data may result in totally different detail and approximation coefficients than the coefficients corresponding to the non-translated data see also Figure 3.1.

Since the retrieval system should also identify translated images with their original, translation invariance of the lifting scheme is a highly desirable property. For the classical wavelet transform a solution to translation invariance is given by the redundant wavelet transform [60], a wavelet transform represented by a filter bank structure without subsampling. As a consequence, the data in all subbands of the filter bank have the same size as the size of the original data. Furthermore, at each resolution level, we have to use zero padding to the filters in order to keep the MRA consistent. Not only is more memory used by the redundant transform, the computing complexity also increases. For the redundant transform computing complexity is $\mathcal{O}(N \log N)$ instead of $\mathcal{O}(N)$ for the FWT.

Whether the redundant transform described is also invariant under reflections and certain rotations depends strongly on the filter structures themselves. Symmetry of the filters is necessary to guarantee the invariance under reflections in the horizontal, vertical and diagonal axis, as well as the invariance under rotations by a multiples of 90 degrees. Symmetry of the filters holds for the Neville filters we are using for this application.

The redundant wavelet transform can also be translated into a redundant lifting scheme. In our case it works out as follows. Instead of subdividing I we copy the image data to both channels in the scheme (I_r, I_b). Next, the scheme is applied on these data, using zero padding for the prediction and update filters at each resolution level. Remark, that for each lifting step in the redundant quincunx lifting scheme we have to store at each resolution level twice as much data as in the non-redundant scheme.

5.1.4 Feature Vectors

After preprocessing all images in the database using the lifting scheme M successive times we arrive at M detail images $I_{d_1}, \ldots, I_{d_M}$. A mathematical 'fingerprint' of these detail images is constructed in order to compare different images by means of these fingerprints. Such identification of an image is realized through a so-called feature vector, a vector that is built up by a set of statistical quantities of each I_{d_l}. An automatic image retrieval system can now be constructed by computing distances between each feature vector and sorting the images on descending distance to the query image. Well-chosen feature vectors are those vectors that have small mutual distances when corresponding to images of the same class and large mutual distances for images of different classes.

The quantities that are used for constructing the feature vectors in our case were first described by Hu [45]. They are based on $(p+q)th$-order central moments $\mu_{pq}(I)$ of a non-discrete image I given by

$$\mu_{pq}(I) = \int_{-\infty}^{\infty}\int_{-\infty}^{\infty} x^p y^q \, I(x+x_c, y+y_c) \, dx \, dy, \qquad (5.7)$$

with the centre of mass

$$x_c = \frac{\int_{-\infty}^{\infty}\int_{-\infty}^{\infty} x I(x,y) \, dx \, dy}{\int_{-\infty}^{\infty}\int_{-\infty}^{\infty} I(x,y) \, dx \, dy} \quad \text{and} \quad y_c = \frac{\int_{-\infty}^{\infty}\int_{-\infty}^{\infty} y I(x,y) \, dx \, dy}{\int_{-\infty}^{\infty}\int_{-\infty}^{\infty} I(x,y) \, dx \, dy}. \qquad (5.8)$$

We observe that the integrals exist since an image is a 2d-function of compact support. For discrete images, based on pixels, the integrals are computed by means of first-order interpolation on the pixels. In his paper Hu derived seven polynomials, based on moments μ_{pq} up to third-order $(p+q=3)$, that attain the same values for an image after translations, rotations and reflections. An example of such an invariant polynomial is given by

$$\mu_{20} + \mu_{02} = \int_{-\infty}^{\infty}\int_{-\infty}^{\infty} (x^2+y^2) \, I(x+x_c, y+y_c) \, dx \, dy, \qquad (5.9)$$

which equals the variance of an image computed from its centre of mass and with its mean subtracted. We observe that the variance of an image object measures the spread of features within the object and therefore, the desired invariances are guaranteed.

Scaling invariance of the moments and the derived polynomials can also be achieved if the moments μ_{pq} in the polynomial invariants are replaced by, for example, $\frac{\mu_{pq}}{\mu_{00}^{\frac{p+q+2}{2}}}$ or $\frac{\mu_{pq}}{(\mu_{20}+\mu_{02})^{\frac{p+q+2}{4}}}$, as shown in [61]. Furthermore, in the same paper the features were normalized in such a way that a multiplication of I with a scalar $\lambda > 0$ results in a multiplication of each feature with the same scalar λ.

Concluding, if we start with an image I, then the lifting scheme generates detail images $I_d^1, \ldots, I_d^M$ using redundant lifting. Computing seven invariants $(F_1, \ldots, F_7)$ of each detail image results in a $7M$-dimensional feature vector

$$\boldsymbol{F}(I) = (F_{1,1}, \ldots, F_{1,7}, \ldots, F_{M,1}, \ldots, F_{M,7}), \tag{5.10}$$

which is not only invariant under the desired affine transformations but also satisfies

$$\boldsymbol{F}(\lambda I) = \lambda \boldsymbol{F}(I). \tag{5.11}$$

The latter property is important if we also intend to build a system, that is invariant under different lighting conditions. A widely used luminosity model is given by an affine transform on the pixel values of an image, i.e., $\widetilde{I}_{k,m} = \lambda I_{k,m} + b$, with $\lambda > 0$ and b a scalar such that $\widetilde{I}$ remains a grey-value image. The detail images are then given by prediction

$$\begin{aligned}(\widetilde{I}_d)_{k,m} &= (\widetilde{I}_b)_{k,m} - (PI_r)_{k,m} \\ &= \lambda(I_b - PI_b)_{k,m} + b(1 - \sum_{(n,l) \in S_{\widetilde{N}}} a_{(n,l)}^{\widetilde{N}}) \\ &= \lambda(I_b - PI_b)_{k,m} = \lambda(I_d)_{k,m},\end{aligned}$$

if and only if the sum of all filter coefficients is normalized to one, which holds for the Neville filters. Since the offset parameter b in the model vanishes in the detail images I_d, differences in lighting result in a multiplication of the feature vector with a scalar

$$\boldsymbol{F}(\lambda I + b) = \boldsymbol{F}(\lambda I) = \lambda \boldsymbol{F}(I), \tag{5.12}$$

according to this model. Lighting invariance can now be realized when feature vectors are compared by means of their mutual angle.

To make the feature vectors more discriminant, after each lifting step we select a relatively small set of detail coefficients to compute the invariants on. For detail image I_d^j this set of coefficients is given by W_j. All other detail coefficients $(I_d^j)_{k,m}$, $(k,m) \notin W_j$, are put to zero. For this newly constructed 'image' I_d^j the moments μ_{pq}^j are computed. However, before calculating the invariants at this resolution level j we have to take into account that the number of coefficients $|W_j|$ may

5.1 Image Retrieval Using Adaptive Lifting

not be a constant as a function of j. Therefore, moments μ_{pq} computed of I_d^j should be normalized first by dividing each of them by $|W_j|$.

For a good representation of the crisp object the set W_j should only consist of locally dominant coefficients, related to the object. A way to achieve this is to compute for each detail coefficient an approximate value for its gradient. This gradient-based information is used to select the coefficients that are likely to be related to the object and not to the background. This is done by means of a threshold value, like

$$(I_d^j)_{k,m} \in W_j \iff |(\nabla I_d^j)_{k,m}| \geq \sqrt{\sum_{n=1}^{N}\sum_{m=1}^{M} |(\nabla I_d^j)_{k,m}|^2 / NM}, \quad (5.13)$$

with N and M the number of rows and columns in I_d^j respectively. The gradients are approximated numerically. The threshold value in the right-hand side of (5.13) is known in the literature as root-mean-square.

5.1.5 Illustration of Image Retrieval

To illustrate the proposed adaptive lifting approach for image retrieval, a synthetic database of 64 images is used. These images can be divided into eight classes, each one consisting of images of one of the objects depicted in Figure 5.5, but subjected to translation over various distances, reflection over various angles, rotation and scaling (zooming in). Moreover, the images are pasted on an arbitrarily chosen wooden background texture of 256×256 pixels. For the simulation each image was used as

Figure 5.5. Object library of eight images of size 128×128.

a query to retrieve the other seven relevant ones.

112   5. Applications of Nonlinear Lifting in Imaging

The effectiveness of our approach (solid line) is shown in Figure 5.6 with both the ideal case (crosses) and the case in which the lifting scheme with a fixed prediction filter was used (dotted line). In this figure, the performance using an eighth-

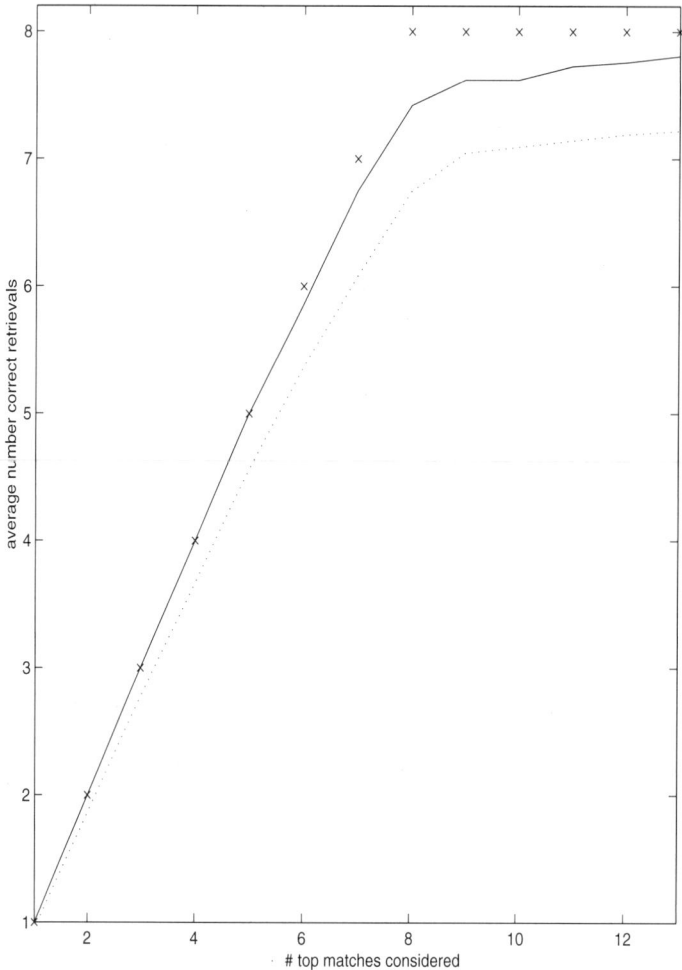

Figure 5.6. Retrieval performance of adapted (solid) and non-adapted (dotted) approaches.

order filter has been depicted since it performed slightly better than lifting with low-order filters. Distances among images are measured by weighted Euclidean distances between the feature vectors. The weighting factors are given by the reverse variances of all distances from the query image to the other images. For simplicity, differences in lighting were not included in this test, so that Euclidean distance measuring is an appropriate tool for measuring similarities.

The average number of retrieved images of the same class as the query image (vertical axis) has been plotted against different numbers of allowed top retrievals. As we can see, retrieval rates increase by 5 to 10% by using an adaptive approach in our test case.

5.2 Adaptive Splitting using Normal Offsets

This section is about new, nonlinear, edge-adaptive multiresolution decompositions for *images*. Data-adaptive, nonlinear lifting is discussed in Chapter 3. Making the (primal and/or dual) lifting steps (i.e., prediction and update operations) depend on the input data may lead to a sparser representation of that input. In many — though not all — applications, sparsity is used as a tool in data compression. In that case, the degree of sparsity is restricted by the condition that the data-adaptive operations have to be invertible, i.e., the lifting operations have to be exactly reconstructed from the output data as they were constructed based on the input data.

Chapter 4 discusses the option of (linear) *grid*-adaptive splitting as a method to reduce instabilities in the multiscale analysis of highly inhomogeneous grids.

This section now introduces a sort of geometrical *data*-adaptive splitting, based on so-called *normal offsets*, not for reasons of stability, but to obtain a sparser data representation, especially in higher dimensions. This focus on the splitting stage, as well as on higher dimensional data, is the main difference between the method discussed in this section and the adaptive lifting discussed in Chapter 3. Data-adaptive splitting, as opposed to grid-adaptivity, makes this method nonlinear. As a matter of fact, it turns out that the grid locations can be considered, in quite a natural way, as part of the data, so they do not need to be stored separately.

We first sketch some of the specific issues in 2-d data processing and explain why these issues have become an important subject in current research on multiscale data representations. Next, we introduce the concept of normal offsets and we illustrate how this concept works in its original context of smooth curve and surface representation. We then turn to the functional setting, and discuss the working of normal offsets when applied for the approximation of piecewise smooth functions.

5.2.1 $2D > (1D)^2$

2-d data processing is essentially more difficult than 1-d data processing, and repeated 1-d techniques on the rows and columns of a (2-d) data matrix are insufficient as a general remedy. In spite of their success in image processing, wavelets are no exception to this rule. This "curse of dimensionality" is situated on both a theoretical and practical level: often, theoretical results in higher dimensions are less far-reaching than their 1-d counterparts, and practical algorithms to achieve the theoretical optimal results are often more difficult to construct.

With respect to the multiscale analysis of 2-d data, we can see at least the following four complications arising from working in more than one dimension. The first three are discussed in other chapters, the last one is the central issue in this section.

Topological – Triangulations. As discussed in Section 2.4, it is generally impossible in two dimensions to construct a neighbourhood system by ordering the observations and defining a neighbour as a 'previous' or 'next' element in that sequence. An alternative route to a neighbourhood system is triangulations. Tensor-product lattices, for instance images, are an exception: there is no need to triangulate data that lie on the corners of a rectangular grid, at least not when it comes to defining a neighbourhood system.

Numerical Stability. As illustrated in Section 4.4.4, Figure 4.7, mixing of scales, as a source of numerical instability, is much more prominent in two dimensions than in one dimension. Major factors in scale mixing are careless lifting operations near the domain boundary or on narrow triangles.

Analytical – Smoothness. In lots of applications, such as image compression or denoising, it is important to construct (piecewise) smooth approximations of given data. Obviously, the smoothness of approximation is determined by the basis functions used for this approximation.

As mentioned in Section 4.3.2, even in one dimension, smoothness of subdivision schemes on irregular point sets proves to be a difficult issue to investigate. Basis functions are no longer translations and dilations of a single father or mother function. Furthermore, Fourier techniques are of little help if the sampling is not periodic. An alternative technique for proving results, called commutation, is adopted in papers that focus on cubic interpolation subdivision [25, 26]. The basic idea is that when a function is the limit of a subdivision process, under some technical conditions, its derivative can be found as the limit of a subdivision process as well and the intermediate values in this subdivision are the divided differences of the original values at corresponding stages [24]. Nevertheless, no general and easy-going technique seems to be available for proofs on smoothness when the samples are irregular.

In higher dimensions, results are much harder to obtain. Multidimensional commutation requires the definition of divided differences on appropriate neighbourhood systems [24]. General results on the smoothness of 2-d subdivision schemes are virtually nonexistent — as far as we know.

Geometrical – Edges = "What" + "Where". The application of wavelets in image analysis, image processing and image compression has been a major factor in the success and proliferation of wavelet methods; it has at least been the most visual and one of the most visible applications. Yet, it can be shown that the classical tensor-product (i.e., separable) 2-d wavelet transform is — at least theoretically — a *suboptimal* representation of typical images. This tensor-product transform is essentially an alternating sequence of wavelet steps on rows of the image and wavelet steps on columns of the image. The corresponding basis functions have squared support, each on a specific scale and specific location in $\mathbb{R}^2$. This squared support turns out to be a limitation in representing (arche)typical images, as explained below.

First we look at wavelets in one dimension. Suppose we have a piecewise smooth function $f(x)$ with a discontinuity or any other singular feature in a point x_0. This singularity could be a discontinuity in the derivative, for instance a cusp, i.e., a point where $f(x)$ is continuous, but its derivative switches sign, i.e., $f'(x_0+) = -f'(x_0-)$. (The prototype of such a feature is the origin in the function $f(x) = |x|$.) In any case, such a singularity affects a wavelet coefficient if the corresponding (dual) wavelet basis function has a support containing x_0. Indeed, the coefficient can be written as the inner product of the basis function with $f(x)$:

$$w_{j,k} = \int_{-\infty}^{\infty} f(x)\tilde{\psi}_{j,k}(x)dx.$$

So, only if $\tilde{\psi}_{j,k}(x)$ is nonzero in the neighbourhood of x_0 does the wavelet coefficient $w_{j,k}$ reflect the singularity. The coefficient is typically large when it describes a singularity and is small otherwise. By construction, the number of basis functions on a given scale whose support contains the singularity is fixed, at least if the number of taps (i.e., nonzero coefficients used in the convolution operation) in all lifting steps at a given scale equals the number of taps in the corresponding lifting step at other scales. In the particular case of an (unbalanced) Haar transform, the basis functions at a given scale have mutually disjunct supports, so only one basis function at each scale gets in touch with a singular point. If the lifting scheme at hand is scale-adaptive, i.e., with possibly different numbers of taps at different scales, the number of basis functions in touch with a singular point may not be a constant, but it is bounded if the number of taps is bounded. Since the total number of basis functions (and corresponding coefficients) increases on finer scales, the relative number of large coefficients, referring to a singular point, vanishes if the resolution level tends to infinity.

As elaborated in [46, Section 6.1.1], this mechanism of increasing sparsity justifies thresholding, i.e., nonlinear approximation of piecewise smooth signals, based on the *largest* wavelet coefficients. The effort in approximating the singularities is comparable or even less than the effort (in terms of approximation error) needed for approximation of the smooth intervals in between two singular points.

More precisely, we have the following result:

Definition 5.1 *A function f is called Lipschitz-α continuous in x if there exist a positive A and a polynomial $p_x(t)$ such that*

$$\forall t \in \mathbb{R} : |f(t) - p_x(t)| \leq A \cdot |t - x|^\alpha. \tag{5.14}$$

A Lipschitz-α continuous function is m times continuously differentiable for all $m \leq \alpha$, but α does not have to be an integer value. In this sense, Lipschitz regularity extends the notion of continuous differentiability to non-integer "derivatives".

Theorem 5.1 *[57, Proposition 9.4] If a square integrable function is piecewise Lipschitz-α continuous, with a finite number of singular points on every bounded* subinterval, and if a wavelet analysis contains at least $p = \lfloor \alpha \rfloor$ (dual) vanishing moments, then the error of an approximation f_n of f with the n largest wavelet coefficients, satisfies

$$\|f_n - f\| := \left(\int |f_n(x) - f(x)|^2 dx \right)^{1/2} = \mathcal{O}(n^{-\alpha}). \tag{5.15}$$

In other words, the approximation error rate is completely determined by the behaviour of f on the smooth subintervals. The singular points have no influence at all on the convergence rate. Such a result could never be obtained with linear methods, such as Fourier or spline methods, where a fixed subset of coefficients (typically the *first*, say the lowest frequencies) is selected in the approximation.

It is remarkable that 1-d wavelets show this nonlinear approximation power without *actively* looking for singularities. The locations (the "where" information) of the singularities follow by looking at the largest coefficients. The approximation in a wavelet basis is mostly nonlinear, but the basis expansion itself is, of course, still linear. An active search for singularities in an *adaptive* so nonlinear, decomposition may be interesting from a practical point of view; the wavelet decomposition, however, provides a tool with a theoretically optimal behaviour.

The picture looks completely different in two dimensions, and the reason is that in two dimensions, singularities are not only located in points, but also — and more often — in lines. These line singularities are (prototypes of) the edges in images. (Edges in real images are often blurred, but we ignore this fact for the moment.) Whereas a point itself has no scale, no dimension, a line has a length. Therefore, it requires more fine-scale than coarse-scale basis functions to capture such an edge. Figure 5.7 illustrates this effect for 2-d tensor-product Haar basis functions on a regular grid. The figure shows shaded squares of different sizes, representing the supports of 2-d Haar basis functions along an edge. The edge is a smooth curve, drawn in full line. It marks the border between two flat (or smooth) regions in the image. One could, for instance, imagine that the pixel values in the right upper corner of the image are all ones, whereas the other side of the edge consists of zeros. It is clear from the figure that there are more fine-scale basis functions than coarse-scale basis functions in contact with the singularity. This observation is due to the fact that a line singularity carries two sorts of information: "what" information (is it a jump, or a cusp, and how large is the discontinuity) and also "where" information. Unlike with point singularities, that "where" information requires intelligent encoding. Tensor-product wavelets are incapable of describing such a line as a whole. Describing that "where" information, i.e., the *geometry*, has been an important topic in recent and current research in multiscale data representation. This topic is often referred to as *(digital) geometry processing*.

The convergence rate of an optimal nonlinear 2-d wavelet approximation with n terms satisfies

$$\|f_n - f\| = \mathcal{O}(n^{-1/2}). \tag{5.16}$$

Thanks to the local support of the wavelet basis functions, this is a faster error decay than the $\mathcal{O}(n^{-1/4})$ of a linear 2-d Fourier approximation. On the other hand, the

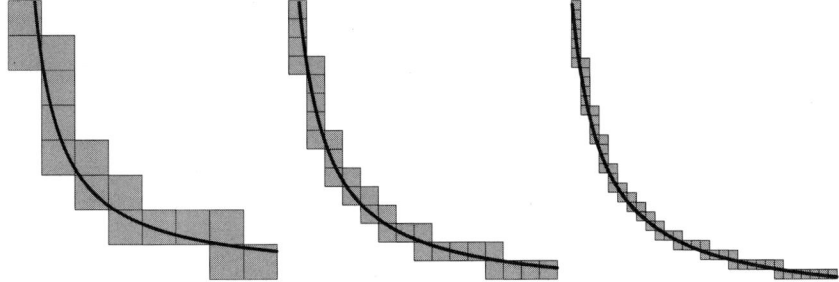

Figure 5.7. Haar wavelet approximation of a 2-d piecewise constant image featuring a smooth 1-d edge singularity. Each shaded square corresponds to the support of a wavelet basis function. At each finer scale, an increasing number of wavelets is necessary to cover (and hence represent) the singularity. This effect does not exist in one dimension and explains the suboptimal performance of wavelets for representing 2-d piecewise smooth functions. Another drawback of tensor-product wavelets is that they approximate the edge curve as a piecewise constant. This explains the "blockiness" of wavelet image approximations.

convergence rate does not depend on the smoothness α of the image outside the edges, nor do the vanishing moments come in at any point. From the approximation theoretic point of view, there seems to be no benefit from exploring 2-d wavelets beyond the Haar case.

Anisotropic Basis Functions. Instead of working with tensor-product basis functions, we could consider more edge-shaped basis functions, i.e., building blocks that are constant or smooth in one direction and oscillating (like wavelets) in the perpendicular direction. Edges in images may be straight or curved (as the one in Figure 5.7), but for simplicity we could start concentrating on straight edges. Since such a decomposition has to provide basis functions along each possible edge of each possible length in each possible direction, the set of functions has to be overcomplete and we need a best basis search algorithm, i.e., an adaptive, active procedure to select an appropriate representation within this overcomplete set for the given image. The result is a basis decomposition where the basis functions are *anisotropic*, i.e., they show different behaviour in different directions. As discussed in Section 5.2.2, such methods exist.

Texture. Tensor-product wavelets are not the ideal tool for analysis and processing of images consisting of large smooth areas, bounded by sharp edges, where the edges themselves are long, smooth curves. Not all images fall within this class, however. The information in some images is dominated by many short edges. Figure 5.8 has an example of such a *textured* image. Texture is characterized by short edges and the crossings of those edges. Those crossings and T-crossings are again point phenomena, for which wavelets are well suited.

Real images may be a mixture of texture and areas with long edges. Even if an image is dominated by long edges, these edges show crossings, so one could speak even then of localized texture. Finally, texture is also a matter of scale: what looks like a short edge in a textured image on a coarse scale may be a long edge between

118   5. Applications of Nonlinear Lifting in Imaging

Figure 5.8. An example of a textured image: the information in this type of image is dominated by many short edges and their crossings. Those crossings are point singularities, for which wavelets are an optimal representation. Edge-oriented decompositions are less appropriate for this type of image.

large smooth areas if one looks at it on a high resolution level. This motivates the need for mixed, adaptive methods that concentrate on edges where necessary and use wavelets to fill in the small details near point singularities or edge crossings.

5.2.2 Some Theoretical Background

This section sketches a situation of some methods in "geometry processing" against a theoretical background. It can easily be skipped in a first reading, or by readers who are mainly interested in the contribution of lifting to the field of geometry processing.

Linear and Nonlinear Approximation. Applications in image processing, such as compression or denoising, can be described within the framework of approximation theory. The goal is to find an approximation g_n of the *target function* f which is a balance between *closeness* to the input y and *smoothness*. Closeness to the input is measured by an error norm, say $\|g_n - y\|$, typically in one of the L_p function spaces, defined on some domain Ω (for instance unit square). Smoothness, on the other hand, is expressed by a smoothness characteristic, a semi-norm, denoted as $|f|_S$. Like a norm, a semi-norm maps elements of a vector space (in our case an L_p function space) onto a non-negative real number; but, unlike a norm, a semi-norm may take a zero value for a nonzero element from the vector space. The approximating g_n comes from a subspace X_n of (in this case) L_p. This subspace is determined by the approximation strategy (linear, nonlinear — so the space is certainly not a classic (linear) vector space in general) and some 'order' n of approximation (in linear approximation this is often the dimension of X_n). The approximation power or capacity is measured by the approximation error, defined as $E_n(f) = \inf_{g \in X_n} \|g - f\|$.

One class of results in approximation theory, already mentioned in Section 4.3.2, are so-called Jackson-type inequalities. They state that the approximation error is bounded by the smoothness semi-norm multiplied by a factor which decays as a function of the order of approximation:

$$E_n(f) \leq C|f|_S n^{-s/d},$$

for some degree of smoothness s and dimension d ($d = 2$ for images). Given a function space L_p and given an approximation strategy, one can impose such an inequality. The inequality then identifies a smoothness semi-norm and — with some further effort — a corresponding approximation space of all functions in L_p with the desired error decay.

For instance, if we want to linearly approximate a periodic function in $L_2([0, 2\pi])$ in a Fourier basis, and if we want an error decay of n^{-s}, the target function has to belong to the Sobolev smoothness space $W^s(L_2([0, 2\pi])) \subset L_2([0, 2\pi])$. Roughly speaking, functions in Sobolev spaces are functions that are in L_2 and whose sth derivative is still in L_2. (For details about the definition of these spaces, we refer to the literature.) Functions in such a space are literally smooth, in the sense that their frequency contents are dominated by low frequencies. The associated Sobolev semi-norm measures the (fractional) sth derivative in the same L_p norm as the approximation error.

Using the same norm to measure error and smoothness is typical in linear approximation. It certainly no longer holds in *nonlinear approximation* [33]. Smoothness of functions with fast L_p error decay in a best n-term wavelet approximation can be measured in a different norm [14]. A typical case is the L_2 error norm, and smoothness measured in L_1. Such a smoothness characteristic favours functions that are smooth except for a limited number of 'transitions' (singularities). In other words, wavelets are the right solution to describe piecewise smooth signals, and smoothness of approximation in this context means *sparsity*. (The trade-off between smoothness and closeness could, therefore, also be seen as a trade-off between low entropy and low error energy.) The smoothness norms naturally arising in the context of best n-term wavelet approximation are Besov norms. Again, we refer to the literature for an exact definition of Besov norms. At this point, we think of Besov norms as a function norm which is small for functions that are smooth, except in some singular points. There exists an equivalent Besov sequence norm, in terms of wavelet coefficients, which essentially measures multiscale sparsity. For well-chosen values of the parameters in a Besov space, wavelets provide the optimal error decay. In two dimensions, for instance, images that live in the Besov space $B^1_{1,1}$ and are viewed in L_2 (i.e., the error is measured in L_2) are best approximated using best n-term wavelet components [7]. (Since wavelet decompositions happen in stable bases, thresholding the coefficients is an easy algorithm to find a best n-term approximation.)

Images ⊂ Besov. This brings us to the question of what sort of images are in $B^1_{1,1}$. Instead of this Besov space, one could also consider the somewhat larger space of functions with bounded variations (BV). This class also includes images with line singularities (i.e., sharp edges) and wavelet decompositions are still near-optimal for images in this space [11, 14]. Neither Besov, nor BV, however, incorporate the *geometrical* information carried by the edges. Images can have large areas of extremely smooth behaviour, separated by long, smooth edges, or they can be textured all over (i.e., with lots of short edges and no global geometry), or they can be a mixture of both, yet the BV smoothness would be almost the same. Wavelets are an ideal tool for textured images, as they are optimal for piecewise smooth 1-d signals. Both textured images and piecewise smooth 1-d signals are also well described by BV or Besov smoothness spaces. The subclass of images with smooth areas and long edges, however, is better described with anisotropic basis functions. This is a typical 2-d problem: in one dimension, isolated singularities are point singularities, with no geometrical information.

Methods in Geometry Processing. One way to deal with the problem of line singularities is, of course, looking for bases or overcomplete representations (frames) that do match with long smooth edges. Contourlets [35] and curvelets [6] are examples of such *anisotropic* basis functions.

Another direction is a so-called *highly nonlinear* approach. This proceeds in two nonlinear steps: a data-adaptive selection of a basis (or a frame) from a collection (library, dictionary) of bases, followed by a best n-term approximation within that basis. Examples here are wedgelets [36, 66], bandelets [55], beamlets [37].

Although the latter transform is nonlinear, the outcome is still a decomposition in a basis (or frame). A third class of methods consists of nonlinear decompositions that cannot be described with basis functions. Most of the methods [1, 3, 15] in this class are related to the idea of adaptive lifting.

The idea of normal offsets [27, 48], presented in Section 5.2.3, can also be seen as an example of nonlinear lifting, but the concept is unique in the sense that the adaptivity does not show up in the lifting steps (prediction, update) itself, but rather in the way this lifting step is used to calculate a 'wavelet' (detail) coefficient. That detail also gets a *geometrical* meaning, and therefore determines *where* exactly a coarse-scale approximation is subdivided. In other words, this is nonlinear lifting with *adaptive splitting*.

Finally, we also mention the possibility of probabilistic geometrical models in Bayesian statistical image processing [49, 46, Chapter 6]. Geometrical priors on Markov random fields, such as the well-known Ising model, may help to link large individual wavelet coefficients into a cluster of coefficients corresponding to an edge.

5.2.3 Normal Offsets

The Principle. The underlying concept of normal offsets is best explained in a 1-d example, illustrated in Figures 5.9 and 5.10. Suppose we want to refine an interpolating polyline approximation of a function. A polyline, already introduced in Section 2.4.1, in the context of triangulations, is a continuous, piecewise linear curve. The example in Figure 5.9 is interpolating, in the sense that the points with discontinuous derivative are evaluations $(x_{j,k}, f(x_{j,k}))$ of an underlying function $f(x)$, which we want to approximate. In order to refine the approximation, we define points at a finer scale as

$$x_{j+1,2k} = x_{j,k}$$
$$x_{j+1,2k+1} = (x_{j,k} + x_{j,k+1})/2,$$

and in the newly inserted points, i.e., the odd indexed points at level $j + 1$, we compute the difference between the approximation and the true function value. This refinement process corresponds to an inverse lifting transform with linear, interpolating prediction and no update. The differences — or offsets — between approximation and true function values are thus details, similar to wavelet coefficients in a lifting scheme with update steps.

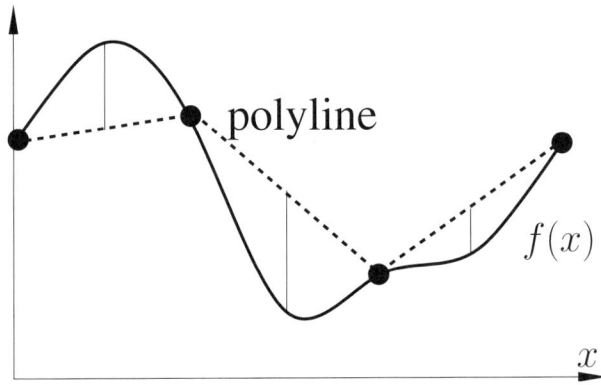

Figure 5.9. Classical refinement step in a multiscale polyline approximation of a function $f(x)$. The current approximation is refined by adding vertical offsets to linear predictions in the middle points of each interval.

Instead of searching for a new point on the curve of $f(x)$ in a fixed point $x_{j+1,2k+1} = (x_{j,k} + x_{j,k+1})/2$, we can also search directions *normal* to the current polyline segments, as illustrated in Figure 5.10. The newly inserted points then have an ordinate $x_{j+1,2k+1}$ which depends on the (slope of) the current approximation. This is a nonlinear refinement.

Smooth Surfaces. The normal refinement procedure was first proposed in the context of smooth surface description [43] for applications in computer graphics. The three-dimensional (3-d) surfaces we are talking about cannot be described as a function $z = f(x, y)$. The curve in Figure 5.11 is a 2-d equivalent. If we want to approximate this curve by general progressive refinement, we need two offsets (detail coefficients) for the localization of each new point. In three dimensions, we need

122   5. Applications of Nonlinear Lifting in Imaging

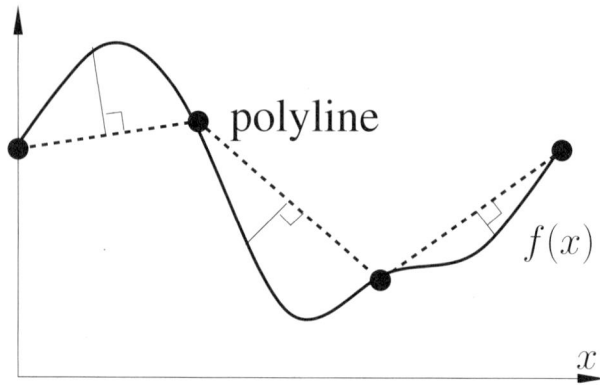

Figure 5.10. Normal refinement step in a multiscale polyline approximation of a function $f(x)$. The current approximation is refined by adding normal offsets to linear predictions in the middle points of each interval.

three detail coefficients. If, however, we decide to look for piercing points in the normal direction, only one coefficient is necessary to indicate how large the offset is in that normal direction.

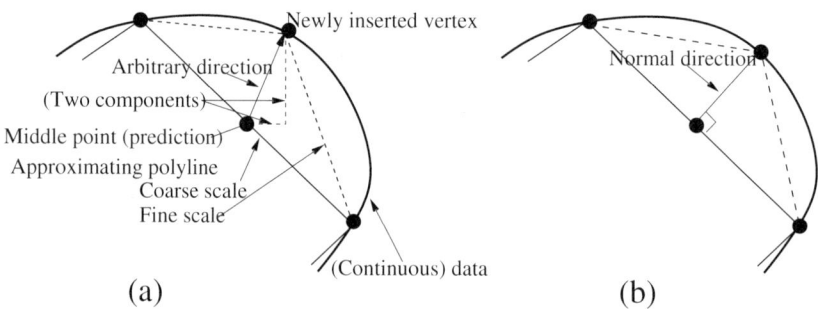

Figure 5.11. General and normal refinement step in a progressive, multiscale approximation of a curve (not necessarily a function). Whereas general refinement requires two coefficients for each new point, normal refinement can do with just one coefficient, since the search direction is fixed.

The Functional Setting: Locating the Edges. The application we have in mind in these sections is not surfaces in computer graphics, but image processing. Images are discretized *functions* on a regular 2-d grid of pixel locations. There is no need to localize these grid points. Grey-scale images can be represented with one pixel value for each location; we do not need three scalars for each pixel in grey-scale images. There is no need to fix a search direction, as the functional setting $z = f(x, y)$ has a natural, fixed search direction, namely that of the third (z) coordinate.

5.2 Adaptive Splitting using Normal Offsets

A second important difference between images and surfaces is edges. The surfaces in computer graphics are generally (though not always) smooth. Images, on the other hand, are dominated by edges.

Figure 5.12 illustrates the behaviour of a multiscale normal offset approximation in the presence of discontinuities. Suppose we have a polyline approximation of a piecewise constant function. The figure shows two successive break points of the polyline, on each side of the discontinuity and the connecting polyline segment. This segment is refined, starting from a prediction in its middle point. If the two existing end points of the segment are sufficiently close to each other and to the location of the singularity (i.e., if we are on a sufficiently fine scale), then the piercing point of the normal direction on the segment with the true function lies *exactly* on the location of the singularity.

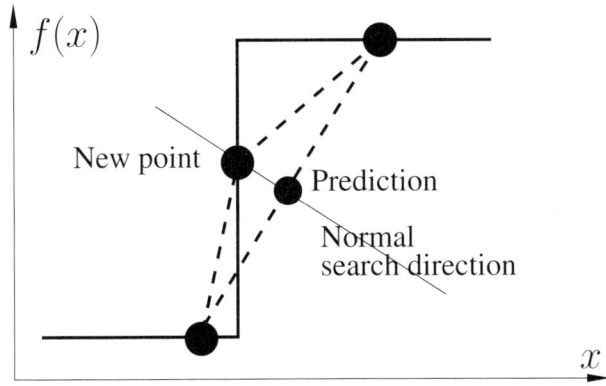

Figure 5.12. Normal offsets near edges: as soon as we are on a sufficiently fine scale, normal offsets point to the exact location of a jump.

Unlike a wavelet analysis, the normal offsets look *actively* for the locations of jumps. Secondly, once the jump has been localized, the error decay continues to be faster than with wavelets [48]. In one dimension this is not really necessary from the asymptotic point of view, as the approximation error will then be dominated by the smooth intervals between the singularities. The real benefit shows up in two dimensions. On the other hand, further analysis [27, 48] reveals that, on smooth functions or smooth parts of piecewise smooth functions, the normal offsets perform as well as wavelets. Normal offsets are, however, a bit slower in the neighbourhood of sharp cusps.

As a matter of fact, normal offsets do not make use of the functional setting of the input data. As a consequence, 'where' information is treated in exactly the same way as 'what' information; the jump on the left of Figure 5.13 shows exactly the same convergence as the two orthogonal cusps on the right. Not making use of the functional setting of input data is, of course, also an example of leaving informa-

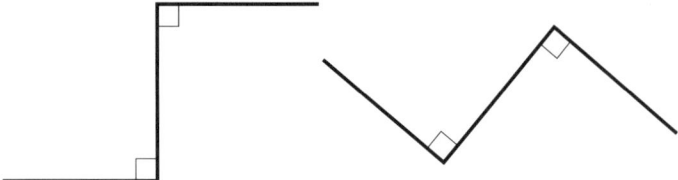

Figure 5.13. A normal polyline does not make any distinction between piecewise constant functions (left) and continuous piecewise linear functions with rectangular cusps (right). As a matter of fact, the notion of function is not that important at all.

tion unexploited. This is already an indication that, for images, we need a mixed approach of classical wavelets and normal offsets.

Normal Offsets in two dimensions: Adaptive Triangulations. The reason for the relatively slow convergence of a nonlinear wavelet approximation in two dimensions, as in Equation 5.16, is two-fold:

1. The non-active localization of the edges by a wavelet decomposition is too slow.
2. As illustrated in Figure 5.7, the ensemble of supports of basis functions on the edge serves as sort of a piecewise constant approximation of that singular line. The approximation power of such a piecewise constant limits the overall rate.

Instead of working with a wavelet basis separable along the traditional x and y axis, we could construct an MRA on a triangular grid as in Figure 5.14(c). If we make that triangulation edge-adaptive, as in Figure 5.14(d), the triangles not only contain the neighbourhood information, but also serve as a piecewise linear approximation of the edges. The edge-locating property of normal offsets, in principle allows us to construct such an adaptive triangulation.

Convergence of Approximation. The convergence analysis [48] of a 2-d normal mesh approximation is based on analyzing the combination of three factors:

1. Fast and active localization of points on the edges. This allows a fast convergence in the xy-plane, in the direction locally perpendicular to the singular line.
2. Piecewise linear approximation by triangulation. This allows a faster-than-wavelets convergence in the xy-plane, in the direction tangential to the singular line.
3. Fast convergence near edge points. Once an edge point is localized, the error near that point has a sufficiently fast decay. Whereas the previous two factors were about approximating the *line* (i.e., the 'where' information) of the singularity, this factor includes the z direction (the dimension of the function values). For the analysis of this factor of convergence, one needs a thorough study of the 1-d behaviour first.

Based on a combination of these three elements, we arrive [48] at the following convergence theorem:

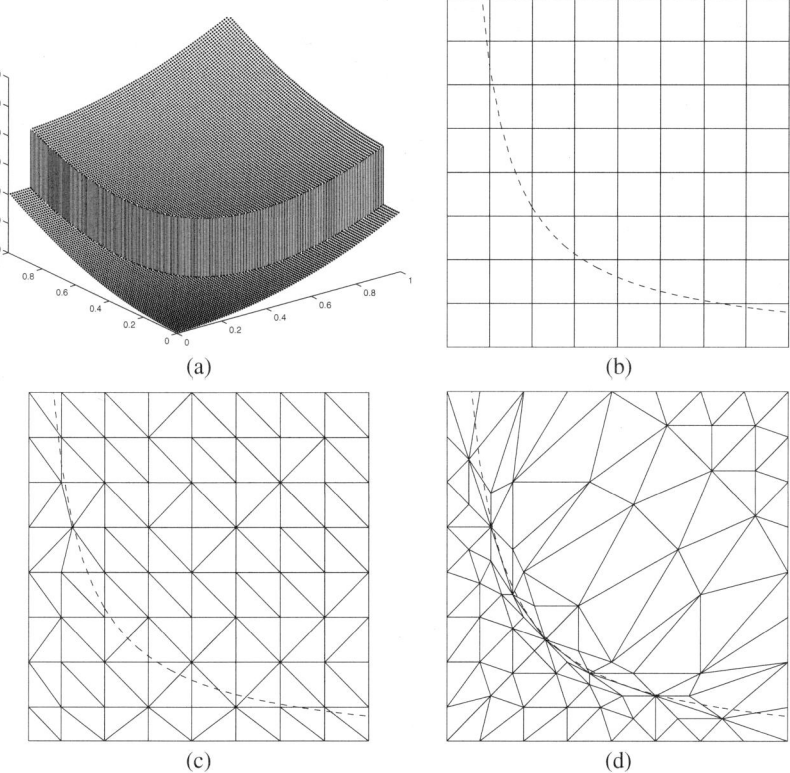

Figure 5.14. Representing the 2-d function $f(x,y) = x^2 + y^2 + I_{\{y>0.1/x\}}$. (a) 3-d mesh plot of the function. (b) A square grid representation such as that used by a tensor-product wavelet transform. The dashed line is the hyperbolic edge in $f(x,y)$. (c) A non-adaptive triangular refinement has the potential of a better approximation, but it does not exploit this potential. (d) The combination of triangulation and adaptivity does the job: triangle sides serve as a piecewise linear approximation of the edges in the image, while others are oriented according to the contour lines in the smooth areas.

Theorem 5.2 *Given a piecewise constant function f defined on $[0,1] \times [0,1]$, or on any compact subset of $\mathbb{R}^2$. Suppose that this function has one singular line (or at least no junctions if there is more than one singular line) and this line is a twice differentiable function with a bounded second derivative.*
Then under mild assumptions *the normal mesh approximation f_n with n nonzero offsets converges in L_2 at a rate of*

$$\|f - f_n\| = \mathcal{O}(n^{-1}).$$

Normal Offsets in Image Analysis. In order to apply the concept of normal offsets to digital images, one needs to consider some technical problems:

126   5. Applications of Nonlinear Lifting in Imaging

1. Without precautions, the adaptive triangulation may run into flipping or crossing triangles and holes, i.e., zones within the function (image) domain that do not belong to any triangle at all. Exception handling procedures, preventing these things from happening, have been worked out [48].
2. The normal piercing point almost never coincides with one of the samples of the discretized curve (or surface). This is a problem, since we want to be able to reconstruct the nearby sample at its original *location* and with its original *value* from a prediction plus the offset. If the normal offset is not able to bring us exactly to the location and value of the sample, we could allow small deviations from a normal search direction, or, equivalently, we could allow an offset leading to the correct value, but not the exact location of the sample. If the deviation of location is smaller than half of the intersample period, this inexact localization can be corrected, but this correction is not possible at sharp transitions, as in Figure 5.15. In order to deal with this problem, we accept a small degree of redundancy: we introduce *two* detail coefficients, as in Figure 5.15. The vector w contains the normal offsets, whereas the vector u contains the local corrections whenever the normal offsets are not sufficient to find both location and height of a sample. This mostly occurs when the normal search direction pierces the true curve in a sharp transition, as in the figure. The proposed procedure inserts two neighbours of this transition in a single step, using two local offsets, $u1$ and $u2$ in the figure.

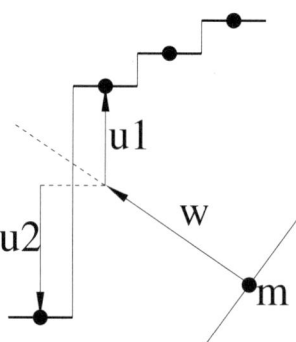

Figure 5.15. Normal offsets for discretized data. Whenever the normal offset w is insufficient to find location *and* value of a new sample, we use two local offsets, $u1$ and $u2$, to fix the values of *two* newly inserted samples. This happens mostly at the location of edges.

In principle, the vectors w and u of normal and local offsets have the same length as the vector f of input observations. The transform is therefore not critically sampled. Nevertheless, the resulting representation is sparser than a wavelet decomposition, as illustrated by the subsequent simulation.

5.2 Adaptive Splitting using Normal Offsets   127

A Simulated Example. We run the normal offset analysis on the data in Figure 5.16. The objective function in this example is piecewise linear and piecewise constant with cusps and jumps as singular points. It was sampled at 2048 equidistant locations, giving an input vector f of 2048 observations. This vector can be represented exactly by 38 nonzero normal offsets together with 100 nonzero local offsets. A decomposition with vertical offsets uses 180 nonzeros, which is more than 100+38. Moreover, most normal offsets can be encoded by the number of positions along the x axis between the location of the prediction and the inserted sample. This is an integer value, so is much easier to compress than general floating-point values. Also, the test signal shows a lot of features on diverse scales. Test signals with only a few edges can be described much more parsimoniously by normal offsets than by vertical offsets. This suggests that a hybrid algorithm, using normal offsets in regions with little features and switching to vertical offsets in textured areas, could combine the benefits from both approaches.

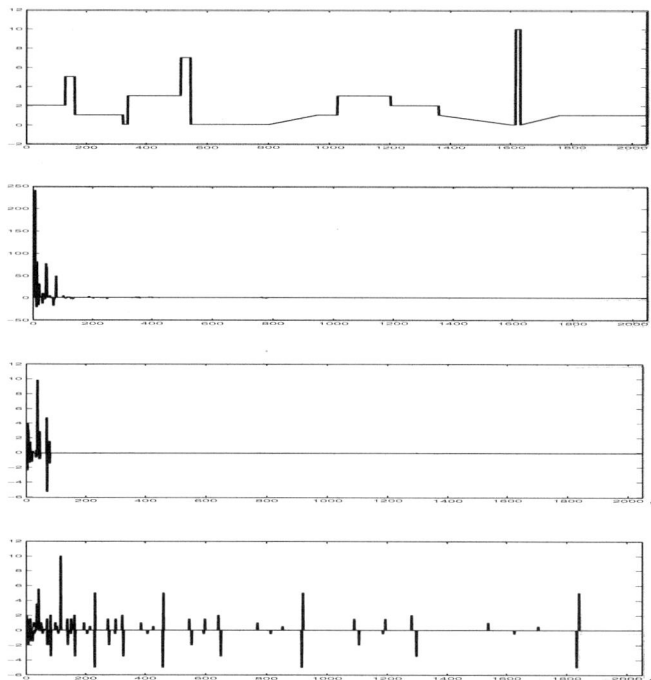

Figure 5.16. Test signal (top), with normal offsets (second) and corresponding local offsets (third), versus vertical offsets (bottom).

References

[1] S. Amat, F. Arandiga, A. Cohen, R. Donat, G. Garcia, and von Oehsen M. Data compression with eno schemes: a case study. *Appl. Comput. Harmon. Anal.*, 11(2):273–288, September 2001.

[2] A. Antoniadis and D.-T. Pham. Wavelet regression for random or irregular design. *Computational Statistics and Data Analysis*, 28(4):333–369, 1998.

[3] F. Arandiga, R. Donat, and Mulet P. Adaptive interpolation of images. *Signal Processing*, 83(2):459–464, February 2003.

[4] T. Cai and L.D. Brown. Wavelet estimation for samples with random uniform design. *Statistics and Probability Letters*, 42:313–321, 1999.

[5] A. Caldéron. Intermediate spaces and interpolation, the complex method. *Stud. Math.*, 24:113–190, 1964.

[6] E. J. Candès and D. L. Donoho. Recovering edges in ill-posed inverse problems optimality of curvelet frames. *Annals of Statistics*, 30:784–842, 2000.

[7] A. Chambolle, R. A. DeVore, N.-Y. Lee, and B. J. Lucier. Nonlinear wavelet image processing: variational problems, compression, and noise removal through wavelet shrinkage. *IEEE Transactions on Image Processing*, 7(3):319–355, March 1998.

[8] C.K. Chui. *An Introduction to Wavelets*. Academic Press, London, 1992.

[9] R. Claypoole, G. M. Davis, W. Sweldens, and R. Baraniuk. Nonlinear wavelet transforms for image coding via lifting. *IEEE Transactions on Image Processing*, 12(12):1449–1459, 2003.

[10] R. L. Claypoole, R.G. Baraniuk, and R. D. Nowak. Lifting construction of non-linear wavelet transforms. In *Proceedings of the IEEE-SP Int. Symp. on Time-Frequency and Time-Scale Analysis*, 1998.

[11] A. Cohen, W. Dahmen, I. Daubechies, and R. DeVore. Harmonic analysis of the space bv. *Revista Matematica Iberoamericana*, 19:235–262, 2003.

[12] A. Cohen and I. Daubechies. A stability criterion for biorthogonal wavelet bases and their related subband coding scheme. *Duke Math. J.*, 68(2):313–335, 1992.

[13] A. Cohen, I. Daubechies, and J. Feauveau. Bi-orthogonal bases of compactly supported wavelets. *Comm. Pure Appl. Math.*, 45:485–560, 1992.

[14] A. Cohen, R. DeVore, P. Petrushev, and H. Xu. Nonlinear approximation and the space $BV(R^2)$. *Amer. J. Math.*, 121:587–628, 1999.

References

[15] A. Cohen, N. Dyn, and B. Matei. Quasilinear subdivision schemes with applications to eno interpolation. *Applied and Computational Harmonic Analysis*, 15:89–116, 2003.

[16] A. Cohen and R. Ryan. *Wavelets and Multiscale Signal Processing*. Chapman & Hall, London, 1995.

[17] L. Cohen. *Time-Frequency Analysis*. Prentice Hall, New Jersey, 1995.

[18] W. Dahmen. Some remarks on multiscale transformations, stability and biorthogonality. In P.J. Laurent, A. Le Mehaut, and L.L. Schumaker, editors, *Wavelets, Images and Surface Fitting*, Curves and Surfaces, II, pages 157–188. AK Peters, Wellesley, MT, 1994.

[19] W. Dahmen. Stability of multiscale transformations. *J. Fourier Anal. Appl.*, 2(4):341–361, 1996.

[20] W. Dahmen, A. Kunoth, and K. Urban. Biorthogonal spline wavelets on the interval: stability and moment conditions. *Appl. Comput. Harmon. Anal.*, 6:132–196, 1999.

[21] I. Daubechies. The wavelet transform, time-frequency localization and signal analysis. *IEEE Transactions on Information Theory*, 36:961–1005, 1990.

[22] I. Daubechies. *Ten Lectures on Wavelets*. CBMS-NSF Regional Conf. Series in Appl. Math., Vol. 61. Society for Industrial and Applied Mathematics, Philadelphia, PA, 1992.

[23] I. Daubechies. Orthonormal bases of compactly supported wavelets II: Variations on a theme. *SIAM J. Math. Anal.*, 24(2):499–519, 1993.

[24] I. Daubechies, I. Guskov, P. Schröder, and W. Sweldens. Wavelets on irregular point sets. *Phil. Trans. R. Soc. Lond. A*, 357:2397–2413, 1999.

[25] I. Daubechies, I. Guskov, and W. Sweldens. Regularity of irregular subdivision. *Constructive Approximation*, 15(3):381–426, 1999.

[26] I. Daubechies, I. Guskov, and W. Sweldens. Commutation for irregular subdivision. *Constructive Approximation*, 17(4):479–514, 2001.

[27] I. Daubechies, O. Runborg, and W. Sweldens. Normal polyline approximation. *Constructive Approximation*, 20(3):399–463, May 2004.

[28] I. Daubechies and W. Sweldens. Factoring wavelet transforms into lifting steps. *J. Fourier Anal. Appl.*, 4(3):245–267, 1998.

[29] V. Delouille, J. Simoens, and R. von Sachs. Smooth design-adapted wavelets for nonparametric stochastic regression. *J. Amer. Statist. Assoc.*, pages 643–658, 2004.

[30] B. Delyon and A. Juditsky. On the computation of wavelet coefficients. *J. of Approx. Theory*, 88:47–79, 1997.

[31] G. Deslauriers and S. Dubuc. Interpolation dyadique. In *Fractales, Dimensions Non-entières et Applications*, pages 44–55. Masson, Paris, 1987.

[32] G. Deslauriers and S. Dubuc. Symmetric iterative interpolation processes. *Constructive Approximation*, 5:49–68, 1989.

[33] R. A. DeVore. Nonlinear approximation. *Acta Numerica*, 7:51–150, 1998.

[34] M. Do, S. Ayer, and M. Vetterli. Invariant image retrieval using wavelet maxima moment. In D.P. Huijsmans and A.W.M. Smeulders, editors, *Proc. Vi-*

sual '99, volume 1614 of *Lecture Notes in Computer Science*, pages 451–458. Springer, 1999.

[35] M. N. Do and M. Vetterli. Contourlets. In G. Welland, editor, *Beyond Wavelets*, pages 83–107. Academic Press, 2003.

[36] D. L. Donoho. Wedgelets: nearly minimax estimation of edges. *Annals of Statistics*, 27(3):859–897, 1999.

[37] D. L. Donoho and X. Huo. Beamlets and multiscale image processing. Technical report, Department of Statistics, Stanford University, 2001.

[38] D. L. Donoho and I. M. Johnstone. Ideal spatial adaptation via wavelet shrinkage. *Biometrika*, 81:425–455, 1994.

[39] D. L. Donoho and I. M. Johnstone. Adapting to unknown smoothness via wavelet shrinkage. *J. Amer. Statist. Assoc.*, 90:1200–1224, 1995.

[40] D. L. Donoho and T.P.Y. Yu. Deslauries-Dubuc: ten years after. In S. Dubuc and G. Deslauriers, editors, *Spline Functions and the Theory of Wavelets*, CRM Proceedings and Lecture Notes. American Mathematical Society, 1999.

[41] J. Goutsias and H.J.A.M. Heijmans. Multiresolution signal decomposition schemes. Part 1: linear and morphological pyramids. *IEEE Transactions on Image Processing*, 9:979–995, 2000.

[42] J. Goutsias and H.J.A.M. Heijmans. Multiresolution signal decomposition schemes. Part 2: morphological wavelets. *IEEE Transactions on Image Processing*, 9:1897–1913, 2000.

[43] I. Guskov, K. Vidimee, W. Sweldens, and P. Schröder. Normal meshes. In *SIGGRAPH 2000 Conference Proceedings*, 2000.

[44] P. Hall and B. A. Turlach. Interpolation methods for nonlinear wavelet regression with irregularly spaced design. *Annals of Statistics*, 25(5):1912–1925, 1997.

[45] M. Hu. Visual pattern recognition by moment invariants. *IRE Trans. Inf. Th.*, 8:179–187, 1962.

[46] M. Jansen. *Noise Reduction by Wavelet Thresholding*, volume 161 of *Lecture Notes in Statistics*. Springer, 2001.

[47] M. Jansen. Wavelet thresholding on non-equispaced data. In D. D. Denison, M. H. Hansen, C. C. Holmes, B. Mallick, and B. Yu, editors, *Nonlinear Estimation and Classification*, volume 171 of *Lecture Notes in Statistics*, pages 257–267. Springer-Verlag, 2002.

[48] M. Jansen, R. Baraniuk, and S. Lavu. Multiscale approximation of piecewise smooth two-dimensional functions using normal triangulated meshes. submitted, 2003.

[49] M. Jansen and A. Bultheel. Empirical Bayes approach to improve wavelet thresholding for image noise reduction. *J. Amer. Statist. Assoc.*, pages 629–639, June 2001.

[50] M. Jansen, G. Nason, and B. Silverman. Scattered data smoothing by empirical Bayesian shrinkage of second-generation wavelet coefficients. In M. A. Unser, A. Aldroubi, and Laine A. F., editors, *Wavelet Applications in Signal*

and Image Processing IX, volume 4478 of *SPIE Proceedings*, pages 87–97, July 2001.

[51] G. Kaiser. *A Friendly Guide to Wavelets*. Birkhäuser, 675 Massachusetts Ave., Cambridge, MA 02139, U.S.A., 1994.

[52] T.H. Koornwinder. *Wavelets: an Elementary Treatment of Theory and Applications*. World Scientific, Singapore, 1993.

[53] A. Kovac and B. W. Silverman. Extending the scope of wavelet regression methods by coefficient-dependent thresholding. *J. Amer. Statist. Assoc.*, 95:172–183, 2000.

[54] J. Kovačević and W. Sweldens. Wavelet families of increasing order in arbitrary dimensions. *IEEE Transactions on Image Processing*, 9:480–496, 1999.

[55] E. Le Pennec and S. Mallat. Sparse geometrical image representations with bandelets. *IEEE Transactions on Image Processing*, 2003.

[56] A. K. Louis, P. Maaß, and A. Rieder. *Wavelets: Theory and Applications*. John Wiley & Sons, 605 Third Avenue, New York, NY 10158-0012, USA, 1997.

[57] S. Mallat. *A Wavelet Tour of Signal Processing*. Academic Press, second edition, 2001.

[58] S. G. Mallat. A theory for multiresolution signal decomposition: The wavelet representation. *IEEE Transactions on Pattern Analysis and Machine Intelligence*, 11(7):674–693, 1989.

[59] Y. Meyer. *Wavelets and Operators*, volume 37 of *Cambridge Studies in Advanced Mathematics*. Cambridge University press, 1992.

[60] G. P. Nason and B. W. Silverman. The stationary wavelet transform and some statistical applications. In A. Antoniadis and G. Oppenheim, editors, *Wavelets and Statistics*, Lecture Notes in Statistics, pages 281–299. Springer, 1995.

[61] P.J. Oonincx and P.M. de Zeeuw. Adaptive lifting for shape-based image retrieval. *Pattern Recognition*, 36:2663–1672, 2003.

[62] P.J. Oonincx and S.J.L. van Eijnhoven. Frames, Riesz systems and MRA in Hilbert spaces. *Indagationes Mathematicae*, 10:369–382, 1999.

[63] P. Oswald. Stable subspace splittings for Sobolev spaces and domain decomposition algorithms. In *Domain Decomposition Methods in Scientific and Engineering Computing (Proc. 7th. Int. Conf. on Domain Decomposition, 1993, Pennsylvania State Univ.)*, volume 180 of *Contemp. Math.*, pages 87–98. Amer. Math. Soc., 1994.

[64] G. Piella. *Adaptive wavelets and their applications to image fusion and compression*. PhD thesis, Department of Mathematics, University of Amsterdam, The Netherlands, 2003.

[65] G. Piella, H.J.A.M. Heijmans, and B. Pesquet-Popescu. Adaptive update lifting with a decision rule based on derivative filters. *IEEE Signal Processing Letters*, 9:329–332, 2002.

[66] J. Romberg, M. Wakin, and R. Baraniuk. Approximation and compression of piecewise smooth images using a wavelet/wedgelet geometric model. In *Proc. IEEE Int. Conf. on Image Proc. — ICIP '03*, 2003.

References

[67] R. Sibson. A vector identity for Dirichlet tessellation. *Mathematical Proceedings of the Cambridge Philosophical Society*, 87:151–155, 1980.

[68] R. Sibson. A brief description of natural neighbour interpolation. In V. Barnett, editor, *Interpolating Multivariate Data*, Lecture Notes in Statistics, pages 21–36. Wiley, 1981. Proceedings of the Conference Entitled "Looking at Multivariate Data", University of Sheffield, U.K., 24–27 March 1980.

[69] J. Simoens and S. Vandewalle. On the stability of wavelet bases in the lifting scheme. TW Report 306, Department of Computer Science, Katholieke Universiteit Leuven, Belgium, 2000.

[70] J. Simoens and S. Vandewalle. A stabilized lifting construction of wavelets on irregular meshes on the interval. *SIAM J. Sci. Comput.*, 24(4):1356–1378, 2003.

[71] R. Stevenson. Locally supported, piecewise polynomial biorthogonal wavelets on nonuniform meshes. *Constructive Approximation*, 19(4):477–508, 2003.

[72] J. Stoer and R. Bulirsch. *Introduction to Numerical Analysis*. Springer, New York, 1980.

[73] G. Strang and T. Nguyen. *Wavelets and Filter Banks*. Wellesley-Cambridge Press, Box 812060, Wellesley MA 02181, fax 617-253-4358, 1996.

[74] W. Sweldens. The lifting scheme: a construction of second generation wavelets. *SIAM J. Math. Anal.*, 29(2):511–546, 1997.

[75] W. Sweldens and P. Schröder. Building your own wavelets at home. In *Wavelets in Computer Graphics*, ACM SIGGRAPH Course Notes, pages 15–87. ACM, 1996.

[76] N.M. Temme. *Special Functions: an Introduction to the Classical Functions of Mathematical Physics*. Wiley, New York, 1996.

[77] G. Uytterhoeven and A. Bultheel. The red-black wavelet transform. TW Report 271, Department of Computer Science, Katholieke Universiteit Leuven, Belgium, December 1997.

[78] E. Vanraes, M. Jansen, and A. Bultheel. Stabilizing wavelet transforms for non-equispaced data smoothing. *Signal Processing*, 82(12):1979–1990, December 2002.

[79] E. Vanraes, J. Maes, and A. Bultheel. Powell-Sabin spline wavelets. *International Journal of Wavelets, Multiresolution and Information Processing.*, 2(1):23–42, 2004.

[80] M. Vetterli and C. Herley. Wavelets and filter banks: theory and design. *IEEE Transactions on Signal Processing*, 40(9):2207–2232, 1992.

[81] P. Wojtaszczyk. *A Mathematical Introduction to Wavelets*. Cambridge University Press, Cambridge, 1997.

[82] J. Xu. Iterative methods by space decomposition and subspace correction. *SIAM Review*, 34(4):581–613, 1992.

[83] H. Yserentant. On the multi-level splitting of finite element spaces. *Numerische Mathematik*, 49:379–412, 1986.

Index

admissibility condition, 3, 92
aliasing, 5, 27
analysis, *see* decomposition
anisotropic basis, 117, 120
average interpolation, 47

B-spline, 13
ball neighbourhood, 53
Banach space, 82
bandelets, 120
basis, 29
basis functions, 4
beamlets, 120
Bernstein estimates, 90
Besov space, 84
biorthogonality, 12
boundary, 54
bounded (operators), 82
boundedly invertible, 82

Calderón's reproducing formula, 1
chirp, 5
commutation, 114
condition number, 81
continuous wavelet transform, 1
− reconstruction formula, 3
contour lines, 125
contourlets, 120
convolution, 2, 5
cosine integral, 5
cubic interpolation, 27
curvelets, 120
cusp, 115

data smoothing, 21, 57, 78

decomposition, 14, 24, 29
Delaunay (triangulation), 55
denoising, *see* data smoothing
dilation equation, *see* two-scale equation
dilation operator, 1
direct estimates, *see* Jackson estimates
discrete wavelet transform, 7
downsampling, 15, 36, 38
dual basis function, 12
dual lifting, 27, 45
dyadic, 21

empirical measure, 43, 50
expansion, 29, 82

fast Fourier transform, 15
fast wavelet transform, 15
father function, *see* scaling function
filter, 5
filterbank, 1, 8, 14, 16, 26, 28, 33
fixed design, 23
forward wavelet transform, *see* decomposition
frames, 85, 120

geometry processing, 116

Hölder, 18
Hölder (continuity), *see* Lipschitz
Haar, 24
Haar measure, 2
Haar wavelet, 3
Hamel basis, 82
Heisenberg inequality, 7

Index

hierarchical basis, 89
highly nonlinear, 120
Hilbert space, 83
homogeneity, 91
homogeneous grid, 91

in-place computation, 28
interpolating prediction, 34, 45
interpolating property, 45
inverse estimates, *see* Bernstein estimates
inverse wavelet transform, 16, 25, 28, 34, 82
inversion formula, 2
irregular grid, *see* non-equispaced data

Jackson estimates, 89, 119

knots, 13

lazy wavelet transform, 38
Legendre polynomials, 83
Lipschitz, 18, 115

mesh, 54
mesh simplification, 56
Mexican hat, 5
mother function, *see* wavelet function
multilevel grid, 41
multiresolution, 8, 29, 43

natural interpolation, 51, 97
non-equidistant data, *see* non-equispaced data
non-equispaced data, 21, 40

orthogonal system, 83

Parseval's equality, 83
Parseval's identity, 2
partition of unity, 19, 92
perfect reconstruction, 8
pilot estimator, 80
polyline, 54, 120
polyphase, 39
prediction, *see* dual lifting

primal lifting, 27, 33
– and stability, 89, 93, 94
pyramid algorithm, 17

quadrature mirror filters, 11
quasi-normalized, 85

random design, 23
rational interpolation, 98
reconstruction, *see* inverse wavelet transform
remeshing, 56, 101
Riesz basis, 8, 85
Riesz constants, 8, 85
Riesz representation theorem, 85

scale mixing, 96
scaling function, 9
scattered data, 23, 40, 56, 57
Schauder basis, 82
Schwarz's inequality, 2
semi-norm, 118
semi-orthogonal, 88
semi-orthogonalization, 95
semi-regular grid, 56
separable basis, 83
separable wavelets, *see* tensor product
series, 29
shrinkage, *see* thresholding
Sibson interpolation, *see* natural interpolation
singular value, 82
splines, 13, 51, 52
stable subspace splitting, 88
subdivision, 33, 77
subdivision connectivity, 56
subsampling, *see* downsampling
synthesis, *see* inverse wavelet transform

tap, 79, 115
tensor-product, 114
texture, 117
thresholding, 21, 78, 84

topological isomorphism, 85
translation operator, 1
triangulation, 54
two-scale equation, 10, 15, 32, 92
two-scale sequence, 10

unbalanced Haar transform, 41
unconditional basis, 83, 84
uniform stability, 87, 88
uniformly continuous, 2
update, *see* primal lifting
upsampling, 16

vanishing moments, 13, 30
– and stability, 93, 94
vertex, 54
Voronoi, 54

wavelet, 9
– filter, 16
wavelet crime, 19, 46, 50
wavelet equation, 10, 32
wavelet function, 3, 9
wavelet transform, 1
wedgelets, 120
window function, 1
windowed Fourier transform, 1

Contact information

Maarten Jansen

TU Eindhoven
Department of Mathematics, H.G. 9.25
P.O. Box 513, NL 5600 MB Eindhoven, the Netherlands
Phone: ++31 (0)40 - 247 4464   Fax: ++31 (0)40 - 246 5995

and

K.U.Leuven
Department of Computer Science
Celestijnenlaan 200A - B 3001 Leuven (Heverlee) Belgium
Phone: ++32 (0)16 - 3 27632   Fax: ++32 (0)16 - 3 27996

E-mail: maarten.jansen@win.tue.nl
WWW: www.cs.kuleuven.ac.be/~maarten/

Patrick Oonincx

Faculty of Military Sciences
Loc. Royal Netherlands Naval College
Department of Navigation Technology
P.O. Box 10000, NL 1780 CA Den Helder, the Netherlands
Phone: ++31 (0)223-657134   Fax: ++31(0)223-657319

E-mail: p.j.oonincx@kim.nl

Maarten Jansen and Patrick Oonincx, July 31, 2004